Solid Oxide Fuel Cell (SOFC) Materials

by R. Saravanan

Developing materials for SOFC applications is one of the key topics in energy research. The book focuses on manganite structured materials, such as doped lanthanum chromites and lanthanum manganites, which have interesting properties: thermal and chemical stability, mixed ionic and electrical conductivity, electrocatalytic activity, magnetocaloric property and colossal magnetoresistance (CMR).

These materials have applications in solid oxide fuel cells, high temperature NOx sensors, hard disk read heads, magnetic sensors and magnetoresistive random access memories.

For the first time, the charge density distributions have been studied in these materials synthesized by high temperature solid state reaction. Charge density analysis is helpful in understanding the physical and chemical properties of materials and in developing optimized structures. The morphological, elemental, optical and magnetic properties of the materials have also been studied.

Solid Oxide Fuel Cell (SOFC) Materials

by

Dr. R. Saravanan, M.Sc., M.Phil., Ph.D.
Associate Professor & Head
Research Centre and PG Department of Physics
The Madura College (Autonomous)
Madurai - 625 011
India

Published by **Materials Research Forum LLC**
Millersville, PA 17551, USA

Published as part of the book series
Materials Research Foundations
Volume 23 (2018)
ISSN 2471-8890 (Print)
ISSN 2471-8904 (Online)

Print ISBN 978-1-945291-50-0
ePDF ISBN 978-1-945291-51-7

Distributed worldwide by

Materials Research Forum LLC
105 Springdale Lane
Millersville, PA 17551
USA
http://www.mrforum.com

Manufactured in the United States of America
10 9 8 7 6 5 4 3 2 1

Table of Contents

Preface

Manganite structured materials have been investigated for the past several decades because of their technological applications in various fields due to their important properties. These materials have mixed valence system with perovskite structure denoted as $R_{1-x}A_xBO_3$, (R-Rare-earth cations (La, Pr, Nd, Sm, Eu, Gd), A- Alkaline earth cations (Ca, Sr, Ba) and B-Transition metal cations (Cr, Mn, Sc, Ni, Fe)). Some of the manganite structured materials are orthochromites, orthomanganites, orthoferrites, orthonickelates and orthoscandates. In this present research work, the manganite structured materials such as doped lanthanum chromites and lanthanum manganites have been analyzed. These doped lanthanum chromite and lanthanum manganite materials have important properties such as thermal and chemical stability, mixed ionic and electrical conductivity, electro catalytic activity, magnetocaloric property and colossal magnetoresistance (CMR). Because of these properties, these materials have been used in solid oxide fuel cells (SOFC), high temperature NO_x sensors, hard disk read heads, magnetic sensors and magnetoresistive random access memories.

The present book mainly focuses on the charge density distribution studies of the manganite structured materials such as doped lanthanum chromites and lanthanum manganites using experimental X-ray diffraction data. In the present book, three series of doped lanthanum chromite samples and two series of doped lanthanum manganite samples have been considered for investigation. Charge density analysis provides the knowledge on the physical and chemical properties of the materials. The charge density distribution studies on prime crystallographic planes provide an understanding of localization/delocalization of charges which is essential for the chromite and manganite materials used in SOFCs. These chromite and manganite samples have been synthesized using the high temperature solid state reaction method. The morphological, elemental, optical and magnetic properties of the materials have also been analyzed.

The study of charge density distribution has been done for the first time, for the synthesized materials.

Chapter 1 gives the objectives of this present research work. It gives detailed information of the materials having manganite structures which include the doped lanthanum chromites and lanthanum manganites. It reviews the literature on undoped and doped lanthanum chromite and manganite materials. It explains the solid state reaction method for the synthesis of the manganite structure materials and gives the synthesis procedure in order to get the required doped lanthanum chromite and manganite materials. It discusses various characterization techniques such as powder X-ray diffraction (XRD), scanning electron microscopy (SEM), energy dispersive X-ray spectroscopy (EDS), UV-visible spectroscopy (UV-vis) and vibrating sample magnetometry (VSM) for the analysis of the structural, morphological, elemental, optical and magnetic properties of the materials. This chapter explains the methodology for the powder XRD profile fitting method. It also explains the concepts and the mathematical formulations of the maximum entropy method (MEM) for the determination of experimental charge density. It further provides the methodologies to estimate the optical band gap and grain size of the materials.

Chapter 2 gives the results obtained from various characterization techniques such as powder XRD, SEM, EDS, UV-vis and VSM for the synthesized lanthanum chromite materials $(La_{0.8}Ca_{0.2})(Cr_{0.9-x}Co_{0.1}Mn_x)O_3$, $(La_{0.8}Ca_{0.2})(Cr_{0.9-x}Co_{0.1}Fe_x)O_3$ & $(La_{0.8}Ca_{0.2})(Cr_{0.9-x}Co_{0.1}Cu_x)O_3$ and manganite materials $La_{1-x}Ca_xMnO_3$ & $La_{1-x}Sr_xMnO_3$. A detailed account of the results of the materials analyzed is given in the subsections.

Section 2.2 gives the raw XRD patterns, the fitted powder XRD profiles and tables for the refined structural parameters for the synthesized chromite and manganite samples. The SEM images, the EDS spectra and tables for elemental compositions have been given in section 2.3. Section 2.4 gives the UV-visible absorption spectra and tables for optical band gap values for the chromite and manganite materials. Magnetization versus magnetic field (M-H) curves and the magnetic parameters have been given in section 2.5. The results from MEM charge density distribution studies are presented as 3D, 2D and 1D electron density maps in section 2.6.

Chapter 3 deals with the analysis of the results obtained from the powder X-ray diffraction, scanning electron microscopy, energy dispersive X-ray spectroscopy, UV-visible spectroscopy and vibrating sample magnetometry for all the synthesized doped lanthanum chromite and manganite materials. The qualitative and quantitative MEM charge density distribution is discussed for all the synthesized samples. An effort has also been made to correlate the structural and magnetic properties.

Chapter 4 presents the conclusion of the results of the reported work.

This book includes research findings which have been previously published by the author as follows;

1.   R. Saravanan, N. Thenmozhi, Yen-Pei Fu, Structural characterization and electron density distribution studies of $(La_{0.8}Ca_{0.2})(Cr_{0.9-x}Co_{0.1}Mn_x)O_3$, *Physica B: Physics of Condensed Matter*, 493, 25-34, (2016) (Elsevier Publication)

2.   R. Saravanan, N. Thenmozhi, Yen-Pei Fu, Preparation and charge density in (Co, Fe)-doped La-Ca-based chromite, *Journal of Electronic Materials*, 45, 4364-4374, (2016) (Springer Publication)

3.   N. Thenmozhi, R. Saravanan, Yen-Pei Fu, Crystal structure and bonding analysis of $(La_{0.8}Ca_{0.2})(Cr_{0.9-x}Co_{0.1}Cu_x)O_3$ ceramics, *Zeitschrift für Naturforschung A- A Journal of*

Physical Sciences, DOI: 10.1515/zna-2016-0474 (De Gruyter Publication) (Published online)

4. N. Thenmozhi, S. Sasikumar, S. Sonai, R. Saravanan, Electronic structure and chemical bonding in $La_{1-x}Sr_xMnO_3$ perovskite ceramics, *Materials Research Express*, 4 (2017) 046103 1-11 DOI.org/10.1088/2053-1591/aa6abf (IOP Publication) (Published online)

5. N. Thenmozhi, R. Saravanan, High temperature synthesis and electronic bonding analysis of Ca-doped $LaMnO_3$ rare-earth manganites, *Rare Metals*, (Springer Publication) (accepted and article in press)

6. N. Thenmozhi, S. Sasikumar, R. Saravanan, Yen-Pei Fu, Study of charge density and crystal structure of co-doped $LaCrO_3$ system, *Mechanics, Materials Science & Engineering*, 9 (2017) DOI: 10.2412/mmse.99.30.568 (Magnolithe GmbH Publication) ISSN: 2412-5954 (Published online)

Chapter 1

Introduction

Abstract

Chapter 1 gives detailed information of materials having manganite structures which include doped lanthanum chromites and lanthanum manganites. It reviews the literature available on undoped and doped lanthanum chromite and manganite materials. It explains the solid state reaction method for the synthesis of the manganite structure materials and gives the synthesis procedure for obtaining the required doped lanthanum chromite and manganite materials. It discusses various characterization techniques such as powder X-ray diffraction (XRD), scanning electron microscopy (SEM), energy dispersive X-ray spectroscopy (EDS), UV-visible spectroscopy (UV-vis) and vibrating sample magnetometry (VSM) for the analysis of the structural, morphological, elemental, optical and magnetic properties of these materials. This chapter also explains the methodology for the powder XRD profile fitting method. It further explains the concepts and the mathematical formulations of the maximum entropy method (MEM) for the determination of experimental charge density. Lastly, it provides the methodologies to estimate the optical band gap and grain size of these materials.

Keywords

Manganite, Lanthanam Chromite Lanthanam Manganite, XRD, UV-Vis, SEM/EDS, VSM

Contents

1.1 Objectives

The main objective of the research work presented in this book is to investigate the growth, physical and X-ray characterization of samples having manganite structures. Manganite structured materials have mixed valence states with perovskite structure denoted as $R_{1-x}A_xBO_3$ (R- Rare-earth cations, A- Alkaline earth cations and B- Transition metal cations). These materials have been widely investigated due to their excellent applications in various fields and their important properties. Orthochromites, orthomanganites, orthoferrites, orthoscandates and orthonickelates are some of the examples for manganite structured materials. The present book analyzes orthochromite and orthomanganite materials, particularly attention is given to doped lanthanum chromites ($LaCrO_3$) and lanthanum manganites ($LaMnO_3$) because of their important applications. Some of the applications are as follows, (i) electrodes and interconnects in solid oxide fuel cells (SOFC), (ii) magnetic sensors, (iii) hard disk read heads and (iv) magnetoresistive random access memory (MRAM).

The following manganite structured materials have been considered for the present investigation.

1. (Co, Mn) doped (La, Ca) based chromites-$(La_{0.8}Ca_{0.2})(Cr_{0.9-x}Co_{0.1}Mn_x)O_3$
2. (Co, Fe) doped (La, Ca) based chromites-$(La_{0.8}Ca_{0.2})(Cr_{0.9-x}Co_{0.1}Fe_x)O_3$
3. (Co, Cu) doped (La, Ca) based chromites-$(La_{0.8}Ca_{0.2})(Cr_{0.9-x}Co_{0.1}Cu_x)O_3$
4. $La_{1-x}Ca_xMnO_3$ manganites
5. $La_{1-x}Sr_xMnO_3$ manganites

and the present research work has the following research tasks.

1. Synthesis of the above series of doped lanthanum chromites and lanthanum manganites using the high temperature solid state reaction method.
2. Study of crystal structural properties by X-ray diffraction (XRD) technique using the Rietveld [Rietveld, 1969] method. The execution of the Rietveld method [Rietveld, 1969] was carried out with the software JANA 2006 [Petříček et. al., 2000].

3. Analysis of the charge density distribution between the atoms in the crystal lattice and bonding features using the maximum entropy method (MEM) [Collins, 1982]. The MEM charge density distribution in the unit cell has been visualized through the software VESTA [Momma and Izumi, 2006]. This study has been done for the first time, for the synthesized manganite structured materials.

4. Analysis of surface morphology and microstructure was done through the images obtained by a scanning electron microscope (SEM).

5. Identification of the elemental compositions of the materials qualitatively and quantitatively using energy dispersive X-ray spectroscopy (EDS).

6. Estimation of optical band gaps (E_g) of the materials by UV-visible absorption spectra using Tauc plot [Wood and Tauc, 1972].

7. Analysis of magnetic properties of the materials by vibrating sample magnetometer (VSM) measurements.

The details of sample preparation, the results obtained from various characterizations and the major conclusion drawn from the results are given in the following chapters.

1.2   Perovskites – An introduction

The research on multifunctional perovskite type oxides ABO_3, is an interesting field for the past five decades due to their functional properties and technological applications. The importance of perovskites is their ability to have different structural distortions due to the possibility of incorporation of nearly every element in the periodic table into its structure [Mitchell et al., 2002]. The properties of perovskites depend primarily on the structure, overlapping of bonds and energy band levels [Coey et al., 1977]. Therefore, for the designing of useful materials, it is essential to predict the structure thoroughly.

In 1839, the mineral perovskite $CaTiO_3$ was discovered by German scientist Gustav Rose, which was named after the Russian scientist Lev Alexeievich Perovsky [Navrotsky et al., 1989]. The perovskites are ceramic materials which have different atomic arrangements. The traditional ABO_3 perovskites consist of small B cations within BO_6 octahedron and the larger A cations are twelve fold coordinated by oxygen. This cubic perovskites have the space group of Pm-$3m$ [Bhalla et al., 2000]. The first study on rare earth perovskites was carried out by Goldschmidt in 1927. Earlier studies reported that the perovskites were mainly of cubic or pseudo-cubic structure. But the recent literature suggests that many materials showed distorted perovskite structures like orthorhombic, rhombohedral and hexagonal structures at room temperature. These orthorhombic, rhombohedral and hexagonal structures have the space group of *Pnma* (or *Pbnm*), *R-3c* and *P6₃cm* respectively [Robert, 1957].

Perovskites with generalized formula ABO_3 consists of three different chemical elements A, B and O in which A and B are metallic cations and O atoms are non-metallic anions. In the cubic ABO_3 structure as shown in figure 1.1, the A cations occupy the eight corners of the cube, B cation is at the center of the cube and O anions are at the middle of the cube's six faces. Hence, the structure of perovskites is in the shape of a cube with an octahedron (BO_6) in it.

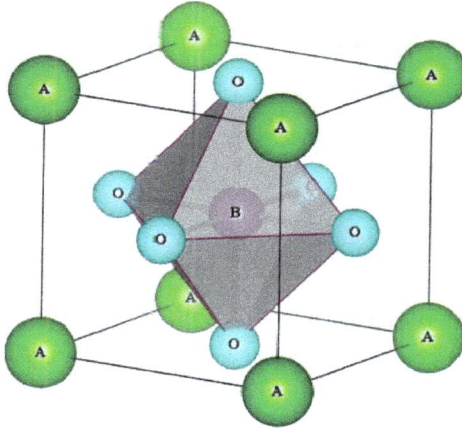

Figure 1.1 *Cubic ABO_3 perovskite structure.*

Due to the mismatch between the sizes of A and B atoms, perovskites are distorted so that B and O atoms move out from their actual positions which lead to an octahedral tilt [Sasaki et. al., 1987]. This octahedral tilt collapses the ideal cubic perovskite frame work and hence alters the physical, electrical, optical and magnetic properties of the perovskites. This distortion can be estimated by the Goldschmidt tolerance factor, $t = (R_A+R_O)/\sqrt{2}(R_B+R_O)$ where R_A, R_B and R_O are the radii of A-site cation, B-site cation and O anion respectively [Goldschmidt, 1926]. If the tolerance factor is close to unity, then the perovskites adopt a cubic structure. For $0.96 < t < 1$, the perovskites are of rhombohedral structure and for $t < 0.96$, the perovskites are of orthorhombic structure. Greater deviations from unity lead to perovskites with hexagonal structure [Pradhan et al., 2013]. Most of the perovskites are distorted and they do not have the ideal cubic structure. In perovskites, the A- site cations are alkali-earths (Ca, Sr, Ba etc.) or rare-

5

earths (La, Pr, Nd, Sm, Eu, Gd etc.) and the B-site cations are transition metals (Cr, Mn, Fe, Sc, Ni, Co etc.). The literature shows that rare-earth orthochromites and orthomanganites have distorted orthorhombic structure with a space group of *Pnma* (Space group no.62) and may also have rhombohedral structure with a space group $R\bar{3}c$ (Space group no.167) [Prado et al., 2013, Coey et al., 1977].

The present research work focuses on samples having manganite structures which include lanthanum chromites ($LaCrO_3$) and lanthanum manganites ($LaMnO_3$) doped with alkaline earth metal ions and transition metal ions.

1.3 Literature review on lanthanum chromites and lanthanum manganites

1.3.1 Crystal structure of La(Cr/Mn)O₃

The basic building block of lanthanum manganite and lanthanum chromite perovskites (ABO_3) is their cubic structure. In the cubic $LaCrO_3$ and $LaMnO_3$ rare earth oxides (figure 1.2), the larger La^{3+} ions (A) are at the corners of the cube, the smaller Cr^{3+}/Mn^{3+} ion is at the center of the cube and the O^{2-} ions are at the face centers of the cube. The face centered oxygen ions coordinate the body centered Cr/Mn ions to form CrO_6/MnO_6 octahedra [Van Aken et al., 2002]. Hence, in the cubic unit cell, La occupies the fractional atomic coordinates (0, 0, 0), Cr/Mn occupies the fractional atomic coordinates (1/2, 1/2, 1/2) and O occupies the fractional atomic coordinates (1/2, 0, 1/2), (1/2, 1/2, 0) and (0, 1/2, 1/2) [Szytuła, 2010]. The undistorted cubic unit cell of La(Cr/Mn)O₃ is shown in figure 1.2.

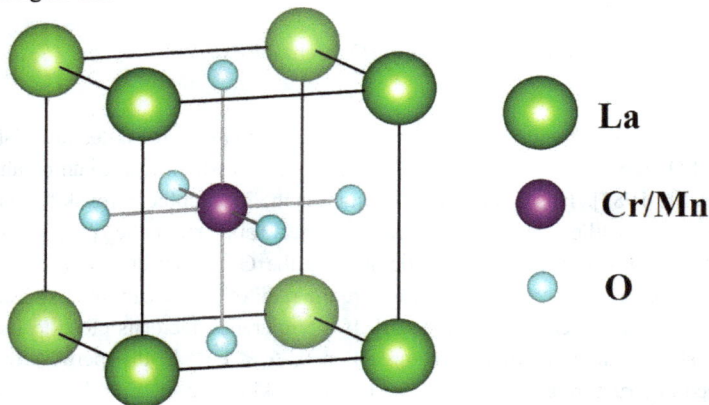

Figure 1.2 Cubic unit cell of La(Cr/Mn)O₃.

Structural distortions can be driven by possible incorporation of alkaline earth ions and transition metal ions at the La site or at the Cr/Mn site or at both sites. These distortions taking place in rare earth perovskites can be explained by three main facts with respect to the ideal cubic structure: (1) tilt of rigid CrO_6/MnO_6 octahedra (2) distortion by polar cation displacements (3) distortion by octahedra (Jahn-teller distortion) [Glazer, 1975]. Out of these three, the most popular distortion taking place in La(Cr/Mn)O_3 perovskites is the distortion of CrO_6/MnO_6 octahedral tilting [Glazer, 2011]. Most of all, the octahedral tilt structure found in perovskites is the orthorhombic structure of space group *Pnma* with tilt system $a^-b^+a^-$ (Glazer's notation) and rhombohedral structure of space group $R\bar{3}c$ with $a^-a^-a^-$ [Abdel-Latif, 2012]. The Goldschmidt tolerance factor depends on the ionic radii of the A-site and B-site cations, which estimates the structural distortion in the perovskites.

Substitution of divalent cations like Ca^{2+}, Sr^{2+}, Ba^{2+} etc., on the La site of La(Cr/Mn)O_3 makes the distortion in CrO_6/MnO_6 octahedra. Hence, Ca^{2+} or Sr^{2+} doped La(Cr/Mn)O_3 structures are either of orthorhombic or rhombohedral structure. Usually, Ca^{2+} doped La(Cr/Mn)O_3 has an orthorhombic structure [Lira-Hernández et al, 2010] whereas Sr^{2+} doped La(Cr/Mn)O_3 has a rhombohedral structure [Corrêa et.al., 2008, Marques et al., 2001]. In the orthorhombic setting of La(Cr/Mn)O_3 (figure 1.3), each unit cell has corner linked octahedra CrO_6/MnO_6 in which the Cr/Mn atoms are at the center and oxygen atoms are at the corners. In the orthorhombic *Pnma* space group structure, the oxygen atoms occupy two non-equivalent positions in the CrO_6/MnO_6 octahedron. The O1 atom is at the apical position whereas the O2 atom is at the equatorial position of the octahedron. The lanthanum atom occupies the space between the octahedra. The chemical unit cell of orthorhombic La(Cr/Mn)O_3 structure is shown in figure 1.3. In the rhombohedral setting of LaMnO_3, the atomic position of the La atom is (0, 0, 0.25), for the Mn atom it is (0, 0, 0) and for the O atom it is (0.5, 0, 0.25). The chemical unit cell of the rhombohedral LaMnO_3 structure is shown in figure 1.4.

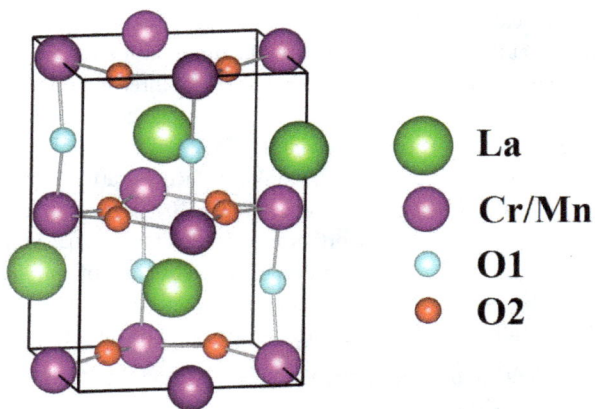

Figure 1.3 *Orthorhombic unit cell of La(Cr/Mn)O₃.*

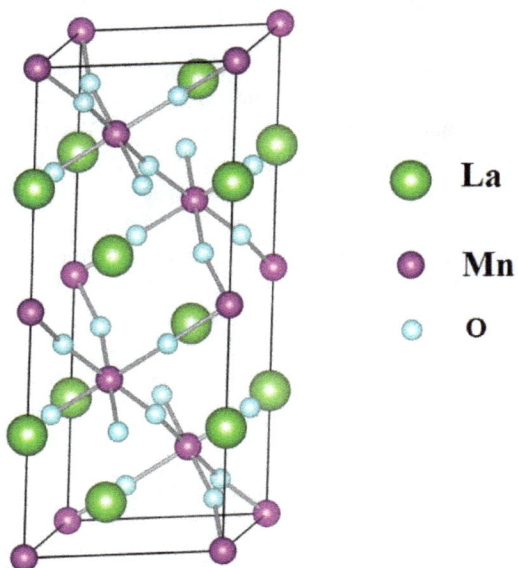

Figure 1.4 *Rhombohedral unit cell of LaMnO₃.*

1.3.2 Properties of La(Cr/Mn)O₃

Lanthanum chromite has the following properties [Nithya et.al., 2012, Setz Luiz et al., 2009],

1. High melting point (~2490°C)

2. High mechanical strength

3. Good electrical conductivity (p-type)

4. High thermal stability

5. Physical and chemical stability in both oxidizing and reducing environment

6. Electrocatalytic activity

7. Mixed ionic and electrical conductivity

8. Refractory nature

9. Magnetocaloric effect

10. Multiferroic property

Like LaCrO₃, lanthanum manganite has the following properties [Abdel-Latif, 2012],

1. Colossal magnetoresistance (CMR)

2. Magnetocaloric property

3. High electrocatalytic activity

4. High electrical conductivity

5. Mixed ionic and electrical conductivity

6. Thermal and chemical stability

7. Multiferroic property

1.3.3 Applications of La(Cr/Mn)O₃

In the early 1960's, perovskite ceramics lanthanum chromites (LaCrO₃) were used for electrode materials in magnetohydrodynamic generators (MHD) [Moulson et al., 2003] since the electrode in MHD requires high electronic conductivity and corrosion resistive property. Because of their ability to withstand high temperatures, high electrical conductivity and electro catalytic activity, lanthanum chromites are used in high temperature NOₓ sensors [West et al., 2005, David et al., 2005], heaters in electric furnaces [Suvorov et al., 2004] and oxidation catalysts for soot combustion [Russo et al., 2005, Ifrah et al., 2007]. Lanthanum chromites are also considered as important materials

9

for electrodes as well as interconnect material in solid oxide fuel cells (SOFC) [Zhu et al., 2004, Petrova et al., 2004]. The ionic conduction behavior of $LaMnO_3$ has led to its use as oxygen sensors [Shu et al., 2009].

1.4 Literature review on doped lanthanum chromites and lanthanum manganites

1.4.1 Properties of doped La(Cr/Mn)O₃

Alkaline earth metal ions (like Ca, Sr, Ba) are generally doped at the La site of lanthanum chromites/manganites and transition metal ions (like Cr, Mn, Fe, Co, Cu etc.) are doped at the Cr/Mn site of lanthanum chromites/manganites [Xifeng et al., 2006]. When Ca^{2+}/Sr^{2+} is doped with $La^{3+}Cr^{3+}O_3$, there is a partial transformation of Cr^{3+} into Cr^{4+} to maintain the overall charge neutrality of the crystalline structure [Larsen et al., 1997] which results in hole doped semiconductivity. In the same way, doping of Ca^{2+}/Sr^{2+} in $La^{3+}Mn^{3+}O_3$, changes the Mn^{3+} ions partially into Mn^{4+} ions and hence $La^{3+}Mn^{3+}O_3$ becomes $La^{3+}_{1-x}(Ca^{2+}_x/Sr^{2+}_x)(Mn^{3+}_{1-x}Mn^{4+}_x)O_3$, which indicates hole doping [Mitchell et al., 1996]. Hence, divalent doping on La(Mn/Cr)O₃, makes the compound a mixed valence system.

In doped manganites, Mn exists in two forms such as Mn^{3+} and Mn^{4+} ions. The Mn^{3+} ion has the electronic configuration as d^4 and in the octahedral setting, the 'd' orbital levels split into three t_{2g} orbitals and two e_g orbitals. Out of the four electrons of the Mn^{3+} ion, three electrons occupy the three t_{2g} orbitals and one electron occupies the doubly degenerate e_g orbital and therefore Mn^{3+} has the configuration as $t_{2g}^3 e_g^1$ (figure 1.5).

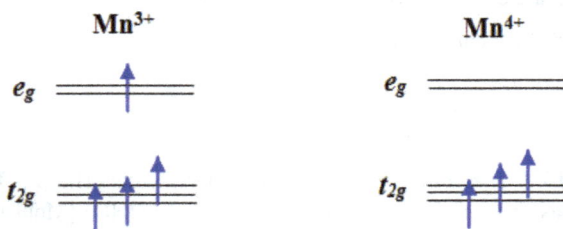

Figure 1.5 Field splitting of manganese (Mn)-5d levels for Mn^{3+} and Mn^{4+}.

The single electron in e_g orbital can act as a mobile charge carrier which is active to the Jahn-Teller (JT) effect. But, the Mn^{4+} ion has the electronic configuration as d^3 which has an empty e_g orbital (no charge carrier) and is inactive to the JT effect [Loa et al., 2001]. The hopping of the itinerant 'e_g' electrons between the two Mn ions (Mn^{3+} and Mn^{4+}) through the oxygen ion induces double exchange interaction [Anderson et al., 1955]. Usually, two different exchange interactions are possible between the Mn ions.

1. t_{2g} (Mn) – 2p (O) - t_{2g} (Mn) → Superexchange interaction

2. e_g (Mn) – 2p (O) - t_{2g} (Mn) → Double exchange interaction

These two interactions taking place between the two Mn ions (Mn^{3+} and Mn^{4+}) in doped manganites explain their magnetic properties. The superexchange interaction between the Mn ions (figure 1.6) with the same valence explains the antiferromagnetic behavior of the compound whereas the double exchange interaction between the Mn ions (figure 1.7) with different valence explains the ferromagnetic behavior of the compound [Pissas et al., 2004]. Hence, divalent ion substitution on La(Mn/Cr)O_3, changes the electronic structure and magnetic properties of the compound.

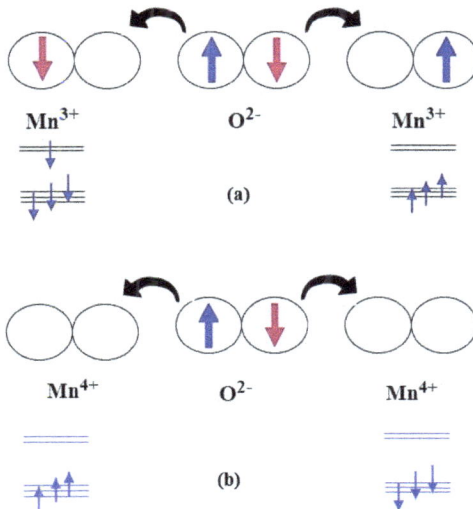

Figure 1.6 Superexchange interactions between (a) Mn^{3+} ions and the O ion. (b) Mn^{4+} ions and the O ion

11

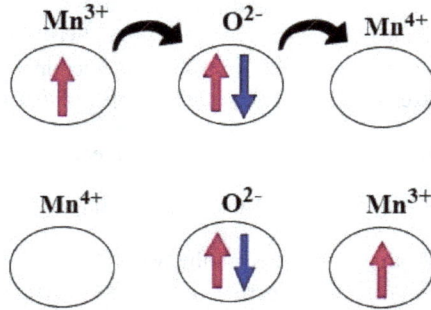

Figure 1.7 Double exchange interactions between two Mn ions and the O ion.

1.4.2 Applications of doped La(Cr/Mn)O$_3$

Lanthanum chromites are considered to be the most suitable materials for interconnectors in SOFC. Lanthanum chromites are p-type electronic semiconductors in oxidizing condition and stable under low oxygen partial pressures. But with low oxygen partial pressures, the p-type conductivity decreases and lanthanum chromites become oxygen deficient. Therefore, to increase the conductivity sufficiently, the electron acceptors are doped with trivalent lanthanum or chromium ions [Setz et al. 2015]. The interconnect materials must have good electrical conductivity, chemical and physical stability in both oxidizing and reducing conditions, good mechanical strength, reasonable thermal conductivity and compatibility with other cell components [Chen et al., 2005]. Hence, the doped (Ca, Sr, Mg, Al, Co, Ni etc.) lanthanum chromites are promising materials for cathodes as well as interconnectors in high temperature solid oxide fuel cells (SOFC). Mg doped lanthanum chromites are used as chemical sensors and as electrode material in magnetohydrodynamic (MHD) generator due to their high catalytic activity [Nadia et al., 2010]. Nano crystalline lanthanum chromites are especially found to be use in catalytic applications [Khetre et al., 2009]. Hayashi et al., have reported that Ca doped LaCrO$_3$ synthesized by the RF magnetron sputtering method, was used to fabricate thin film heaters [Hayashi et al., 2002]. Doped chromites are also used in metallic interconnects as protective coatings [Hilpert et al., 1996].

Lanthanum manganites doped with divalent ions have been studied over the past several decades due to their interesting properties like magnetocaloric and magnetoresistive effects. Sr and Ca doped lanthanum manganites exhibit colossal magnetoresistance property (CMR) and can be used in magnetic sensors and hard disk read heads [Balcells et al., 1996, Daengsakul et al., 2009, Michael, 2000]. They have also been used as catalysts [Bella et al., 2000, Rezlescu et al., 2013]. Strontium doped lanthanum manganites (LSMO) change their property with respect to their spin orientation and hence these manganite perovskites can be used in spintronic devices [Mara et al., 2001] like magnetoresistive random access memory [Jain et al., 2006]. Because of their high T_C (about 360 K) and high magnetic moment, LSMO can be used for hyperthermia applications [Rashid et al., 2013].

Sr and Ca doped lanthanum manganites have been used as the major materials for cathodes used in solid oxide fuel cells (SOFC) because of their high electrical conductivity [Brichzin et al., 2000, Nikolina et al., 2005]. SOFC offers a clean, pollution-free technology to generate electricity. Advantages of SOFC's are their high efficiency, long term stability, fuel flexibility, low emissions and relatively low cost [Michael et al., 2008]. The schematic diagram of SOFC is given in figure 1.8. SOFCs are designed both in tubular and planar form as shown in figure 1.9 and figure 1.10 respectively. Stationary power generation systems of 1 to 25 kW size have been fabricated using planar SOFCs. Planar SOFCs are very attractive for transportation and military applications such as ship service power and ship propulsion, army ground vehicles' auxiliary power units and propulsion, mobile power generators etc. By using tubular SOFCs, power generation systems up to 250 kW size have been produced. These tubular and planar SOFCs are ideal power generation systems-reliable, environmental friendly and fuel conserving [Singhal, 2002].

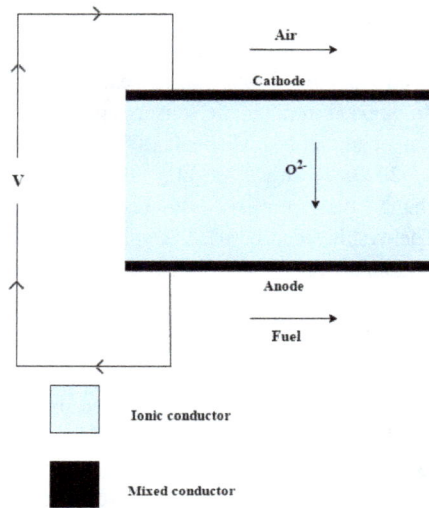

Figure 1.8 *Schematic diagram of solid oxide fuel cell.*

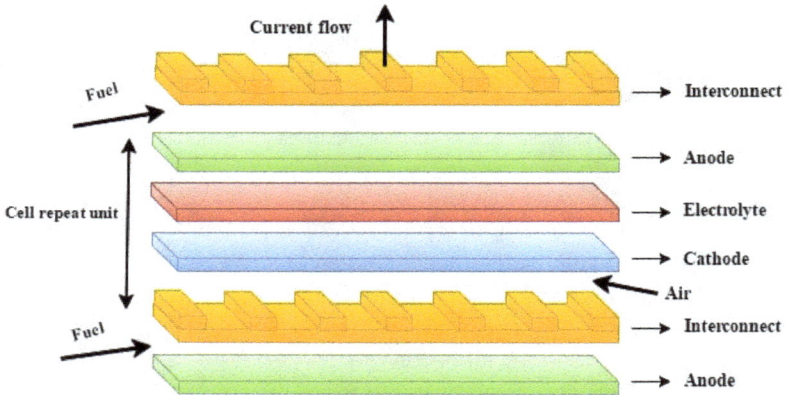

Figure 1.9 *Schematic diagram of planar solid oxide fuel cell.*

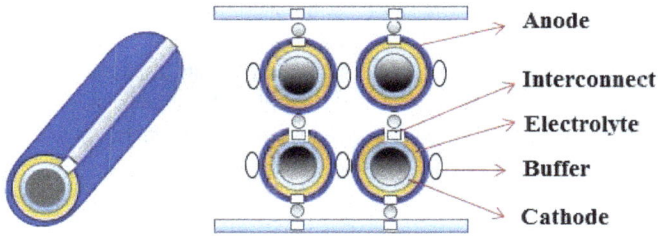

Figure 1.10 Schematic diagram of tubular solid oxide fuel cell.

1.5 Preparation of manganite structure materials

1.5.1 Solid state reaction method

The literature shows many preparation techniques to synthesize doped lanthanum chromites and lanthanum manganites which includes the microwave combustion synthesis [Athawale et al., 2011], the Pechini method [Masashi Mori et al., 2002] the solid state reaction method [Tyson, 1996], the solution combustion method [Bellakki et al., 2010], the citrate sol-gel method [Carlos et al., 2006], the sol-gel method, the soft-chemical method [Reddy et al., 2013], the microwave-assisted hydrothermal method [Rizzuti et al., 2009], the polymeric precursor method [Rabeloa et al., 2011], etc. But to synthesize ceramic materials, the solid state reaction method [Narayan et al., 2009] is favored.

In the solid state reaction method, stoichiometric ratios of the precursor powders should be mixed thoroughly and then the samples should be heat treated with intermediate grinding. During the solid state reaction synthesis, the sample undergoes four main processes which include calcination, grinding, pelletization and sintering [Narayan et al., 2009]. Initially, the precursor powders are calcined at a temperature, lesser than the melting point of the final product material. Then, thorough grinding is needed to get homogenous mixture of reactant powders. To increase the reaction rate between the particles, pelletizing the powders is essential. Sintering of pelletized powder samples should be done at high temperature (lesser than the melting point of the product material) and hence the crucibles (containers of reaction samples) must be able to withstand high temperatures. During sintering, atomic diffusion occurs so that the particles fuse together to form the final product. The visible sign of sintering is the shrinkage of the material.

The advantage of the solid state reaction technique is that the high temperature induces high rate of diffusion and all the samples of varying concentrations are prepared under the same environment. Moreover, there is no need for solvents and no purification is needed after preparation. In the present work, doped lanthanum chromites and lanthanum manganites were synthesized using the high temperature solid state reaction technique.

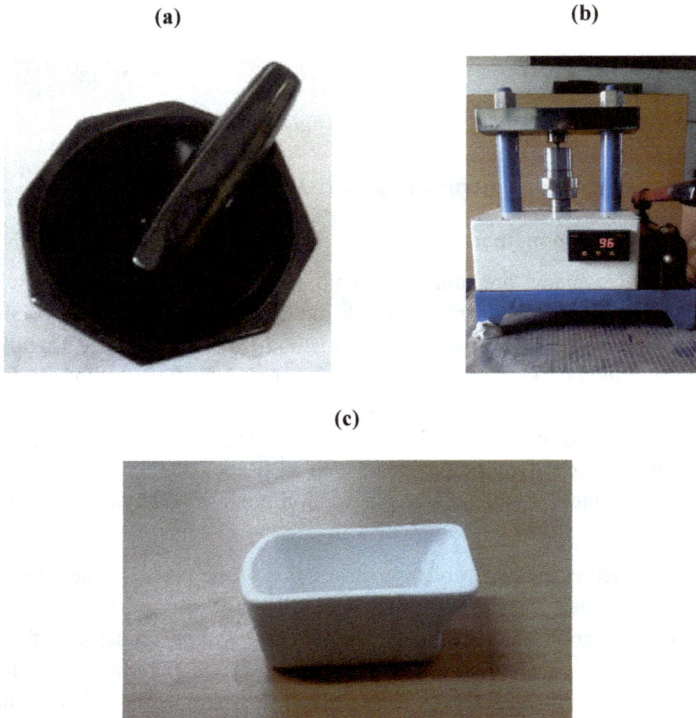

(a)

(b)

(c)

Figure 1.11 Photographs of experimental tools *(a)* Agate mortar and pestle, *(b)* Pelletizer and *(c)* Alumina crucible.

In our work, the raw materials were weighed using the electronic balance model MK 200E which has a readability accuracy of 0.001 gm. For mixing and grinding, agate mortar and pestle was initially used. For pelletizing the samples, a pelletizer which uses a maximum hydraulic pressure of 100 MPa has been used. Two different dies with 1.0 cm and 1.2 cm diameter were used to get the pellets of the prepared materials. For calcinations and sintering, a tubular furnace was used. This furnace has a working temperature of up to 1600 °C. This tubular furnace has a programmable heating rate of 1 °C/min to 5 °C/min and has 1 °C accuracy of dwell temperature. The photographs of the experimental tools used and the tubular furnace are presented in figures 1.11 and 1.12.

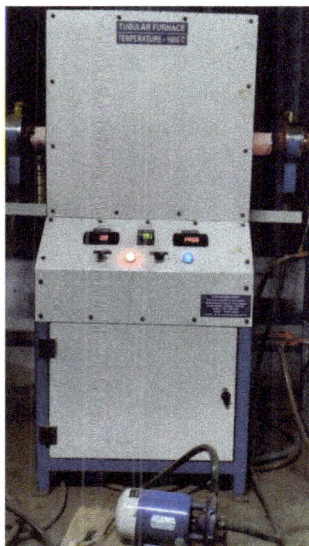

Figure 1.12 Tubular furnace used for calcination and sintering purposes.

1.5.2 Synthesis

1.5.2.1 (Co, Mn) doped (La, Ca) based chromites - $(La_{0.8}Ca_{0.2})(Cr_{0.9-x}Co_{0.1}Mn_x)O_3$

To synthesize the co-doped lanthanum chromites $(La_{0.8}Ca_{0.2})(Cr_{0.9-x}Co_{0.1}Mn_x)O_3$ (x=0.03, 0.06, 0.09 and 0.12), the starting materials lanthanum oxide $(La_2O_3, 99.99\%)$, chromium oxide (Cr_2O_3, 99.99%), cobalt oxide (Co_3O_4, 99.99%), manganese oxide (MnO_2, 99.99%) and calcium carbonate ($CaCO_3$, 99.99%) were used. Stoichiometric quantities of these carbonate and oxide powders were mixed according to the following equation.

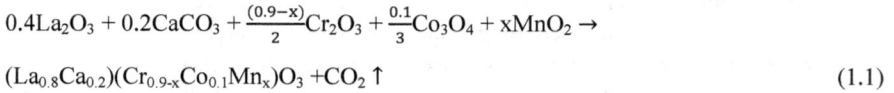

$$0.4La_2O_3 + 0.2CaCO_3 + \frac{(0.9-x)}{2}Cr_2O_3 + \frac{0.1}{3}Co_3O_4 + xMnO_2 \rightarrow$$

$$(La_{0.8}Ca_{0.2})(Cr_{0.9-x}Co_{0.1}Mn_x)O_3 + CO_2 \uparrow \qquad (1.1)$$

Stoichiometric quantities of precursor materials were taken according to their molecular weights given in table 1.1 and used according to equation (1.1). Table 1.2 gives the actual quantities of carbonate and oxide powders that were used for the synthesis of (Co, Mn) doped (La, Ca) based chromites. These quantities of reagents were then stirred with distilled water for 12 h. After drying, the powders were thoroughly ground using an agate mortar and then calcined in air at 1000 °C for 4 h. Then, the calcined powder samples were pelletized. These pellets were again dry pressed at 100 MPa. Then, the pellets were sintered at 1500 °C in air for 6 h with a heating rate of 5°C/min. The synthesized $(La_{0.8}Ca_{0.2})(Cr_{0.9-x}Co_{0.1}Mn_x)O_3$ (x=0.03, 0.06, 0.09 and 0.12) pellets are shown in figure 1.13.

Table 1.1 *Molecular weights of chemicals used.*

Chemicals	Molecular weight (gm/mol)
La_2O_3	325.81
$CaCO_3$	100.08
$SrCO_3$	147.63
Cr_2O_3	151.99
Co_3O_4	240.80
MnO_2	86.93
Fe_2O_3	159.69
CuO	79.54

Table 1.2 *Quantities of starting materials to synthesis (Co, Mn) doped (La, Ca) based chromites.*

Concentration (x)	La_2O_3 (gm)	$CaCO_3$ (gm)	Cr_2O_3 (gm)	Co_3O_4 (gm)	MnO_2 (gm)
0.03	5.212	0.801	2.645	0.321	0.104
0.06	5.212	0.801	2.553	0.321	0.209
0.09	5.212	0.801	2.462	0.321	0.313
0.12	5.212	0.801	2.370	0.321	0.417

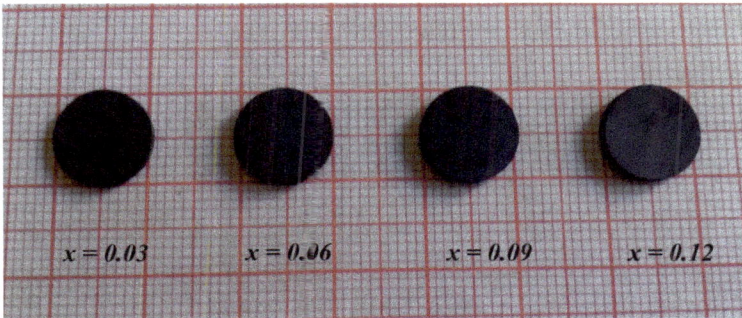

Figure 1.13 *The synthesized samples of $(La_{0.8}Ca_{0.2})(Cr_{0.9-x}Co_{0.1}Mn_x)O_3$, x=0.03, 0.06, 0.09 and 0.12.*

1.5.2.2 (Co, Fe) doped (La, Ca) based chromites - $(La_{0.8}Ca_{0.2})(Cr_{0.9-x}Co_{0.1}Fe_x)O_3$

The transition metal doped perovskite compound, $(La_{0.8}Ca_{0.2})$ $(Cr_{0.9-x}Co_{0.1}Fe_x)$ O_3 (x=0.03, 0.06, 0.09, 0.12) was synthesized by mixing the powders of lanthanum oxide (La_2O_3, 99.99%), chromium oxide (Cr_2O_3, 99.99%), cobalt oxide (Co_3O_4, 99.99%), ferric oxide (Fe_2O_3, 99.99%) and calcium carbonate ($CaCO_3$, 99.99%) in stoichiometric ratio according to the following equation.

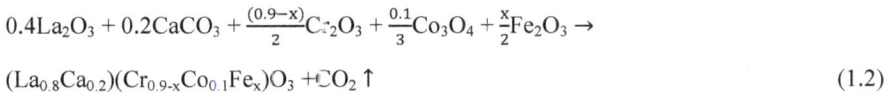

$$0.4La_2O_3 + 0.2CaCO_3 + \frac{(0.9-x)}{2}Cr_2O_3 + \frac{0.1}{3}Co_3O_4 + \frac{x}{2}Fe_2O_3 \rightarrow$$

$$(La_{0.8}Ca_{0.2})(Cr_{0.9-x}Co_{0.1}Fe_x)O_3 + CO_2 \uparrow \qquad (1.2)$$

The appropriate quantities of starting materials were calculated using equation (1.2) and the molecular weights given in table 1.1. To synthesize (Co, Fe) doped (La, Ca) based chromites, the required quantities of starting materials as given in the table 1.3 were used. Initially, stoichiometric quantities of the starting materials were stirred with distilled water for 12 h. Then the powders were dried and ground well using an agate mortar. These ground powders were calcined in air at 1000 °C for 4 h. The calcined powders were then pelletized and the pellets were dry pressed at 100 MPa. Finally, the pressed pellets were sintered at 1500 °C in air for 6 h.

The synthesized $(La_{0.8}Ca_{0.2})(Cr_{0.9-x}Co_{0.1}Fe_x)O_3$ (x=0.03, 0.06, 0.09 and 0.12) pellets are shown in figure 1.14.

Table 1.3 *Quantities of starting materials to synthesis (Co, Fe) doped (La, Ca) based chromites.*

Concentration (x)	La_2O_3 (gm)	$CaCO_3$ (gm)	Cr_2O_3 (gm)	Co_3O_4 (gm)	Fe_2O_3 (gm)
0.03	5.212	0.801	2.645	0.321	0.096
0.06	5.212	0.801	2.553	0.321	0.192
0.09	5.212	0.801	2.462	0.321	0.287
0.12	5.212	0.801	2.370	0.321	0.383

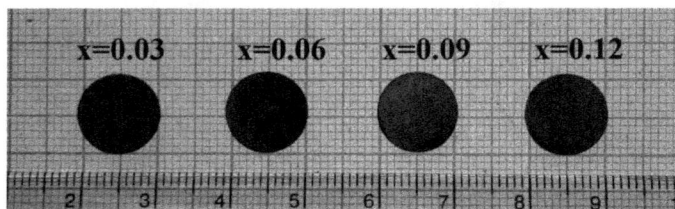

Figure 1.14 *The synthesized samples of $(La_{0.8}Ca_{0.2})(Cr_{0.9-x}Co_{0.1}Fe_x)O_3$, x=0.03, 0.06, 0.09 and 0.12.*

1.5.2.3 (Co, Cu) doped (La, Ca) based chromites – $(La_{0.8}Ca_{0.2})(Cr_{0.9-x}Co_{0.1}Cu_x)O_3$

The co-doped lanthanum chromite samples $(La_{0.8}Ca_{0.2})(Cr_{0.9-x}Co_{0.1}Cu_x)O_3$ (x=0.00, 0.03 and 0.12) were synthesized using the solid state reaction process. High-purity of lanthanum oxide (La_2O_3, 99.99%), chromium oxide (Cr_2O_3, 99.99%), cobalt oxide (Co_3O_4, 99.99%), copper oxide (CuO, 99.99%) and calcium carbonate ($CaCO_3$, 99.99%)

were used as starting materials. Stoichiometric quantities of these materials were mixed according to the following equation.

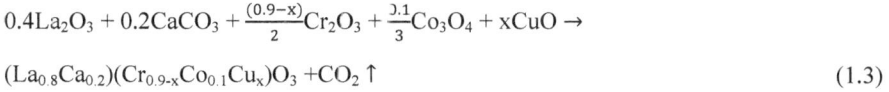

$$0.4La_2O_3 + 0.2CaCO_3 + \frac{(0.9-x)}{2}Cr_2O_3 + \frac{0.1}{3}Co_3O_4 + xCuO \rightarrow$$

$$(La_{0.8}Ca_{0.2})(Cr_{0.9-x}Co_{0.1}Cu_x)O_3 + CO_2 \uparrow \tag{1.3}$$

Table 1.4 gives the quantities of the starting materials for the synthesis of (Co, Cu) doped (La, Ca) based chromites. As the first step in the synthesis process, the starting materials were stirred with distilled water for 12 hours. Then, these powders were dried and ground thoroughly. Thereafter, the ground powders were calcined in air at 1000 °C for 4 h. Then the resultant powders were pelletized and dry pressed at 100 MPa. Then the pellets were sintered at 1500 °C in air for 6 h. The synthesized chromite pellets of $(La_{0.8}Ca_{0.2})(Cr_{0.9-x}Co_{0.1}Cu_x)O_3$ (x=0.00, 0.03 and 0.12) are shown in figure 1.15.

Table 1.4 *Quantities of starting materials to synthesis (Co, Cu) doped (La, Ca) based chromites.*

Concentration (x)	La₂O₃ (gm)	CaCO₃ (gm)	Cr₂O₃ (gm)	Co₃O₄ (gm)	CuO (gm)
0.00	5.212	0.801	2.735	0.321	0.000
0.03	5.212	0.801	2.645	0.321	0.095
0.12	5.212	0.801	2.370	0.321	0.382

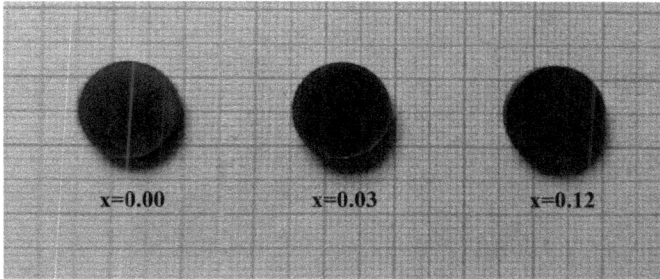

Figure 1.15 *The synthesized samples of $(La_{0.8}Ca_{0.2})(Cr_{0.9-x}Co_{0.1}Cu_x)O_3$, x=0.00, 0.03 and 0.12.*

1.5.2.4 La$_{1-x}$Ca$_x$MnO$_3$ manganites

Calcium doped lanthanum manganite polycrystalline samples were synthesized using high purity lanthanum oxide (La$_2$O$_3$, 99.99%), calcium carbonate (CaCO$_3$, 99.99%) and manganese oxide (MnO$_2$, 99.99%) as starting materials. Stoichiometric quantities of these powders were taken according to the following equation.

$$\frac{(1-x)}{2} La_2O_3 + x\ CaCO_3 + MnO_2 \rightarrow La_{1-x}Ca_xMnO_3 + CO_2 \uparrow \qquad (1.4)$$

Using the molecular weights of the starting materials and the equation (1.4), the quantities of starting materials were calculated and given in table 1.5. These oxide powders were mixed and ground for 2 h using an agate mortar. These ground powders were then calcined in a tubular furnace at 1300 °C for 5 h for homogeneity. Then, the calcined powders were ground again using an agate mortar for 3 h and further calcined at 1400 °C for 12 h. Thereafter, the powders were ground again with an agate mortar for 3 h. These powders were then pelletized and the resultant pellets were further sintered at 1450 °C for 15 h in a tubular furnace to get the required samples. The prepared La$_{1-x}$Ca$_x$MnO$_3$ (x=0.1, 0.2, 0.3, 0.4 and 0.5) manganite pellets are shown in figure 1.16.

Table 1.5 Quantities of starting materials to synthesis Ca doped lanthanum manganites.

Concentration (x)	La$_2$O$_3$ (gm)	CaCO$_3$ (gm)	MnO$_2$ (gm)
0.1	2.443	0.167	1.449
0.2	2.172	0.334	1.449
0.3	1.901	0.501	1.449
0.4	1.629	0.667	1.449
0.5	1.357	0.834	1.449

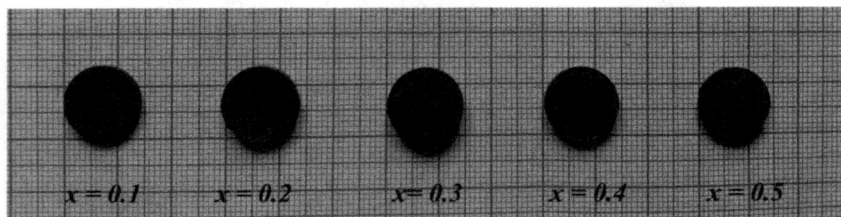

Figure 1.16 The synthesized samples of La$_{1-x}$Ca$_x$MnO$_3$, x=0.1, 0.2, 0.3, 0.4 and 0.5.

1.5.2.5 $La_{1-x}Sr_xMnO_3$ manganites

The starting materials for the synthesis of strontium doped lanthanum manganite were lanthanum oxide (La_2O_3, Alfa Aesar 99.99%), strontium carbonate ($SrCO_3$, Alfa Aesar 99.99%) and manganese oxide (MnO_2, Alfa Aesar 99.99%). Proper weight ratios of these oxide and carbonate powders were used according following equation.

$$\frac{(1-x)}{2} La_2O_3 + x\ SrCO_3 + MnO_2 \rightarrow La_{1-x}Sr_xMnO_3 + CO_2 \uparrow \qquad (1.5)$$

The quantities of starting materials used are given in table 1.6. These weighed powders were mixed thoroughly and ground for 3 h using an agate mortar. Then, these ground powders were calcined at 1250 °C for 8 h. Then the calcined powders were ground again for 5 h using agate mortar and further sintered at 1400°C for 12 h. These sintered powders were ground once again for 3 h using an agate mortar. The resultant powders were pressed into pellets and sintered at 1450 °C for 15 h. Then, the final sintered powders were ground again for further characterizations. The synthesized pellets of $La_{1-x}Sr_xMnO_3$ (x=0.3, 0.4 and 0.5) manganites are shown in figure 1.17.

Table 1.6 *Quantities of starting materials to synthesis Sr doped lanthanum manganites.*

Concentration (x)	La_2O_3 (gm)	$SrCO_3$ (gm)	MnO_2 (gm)
0.3	2.281	0.886	1.739
0.4	1.955	1.181	1.739
0.5	1.629	1.476	1.739

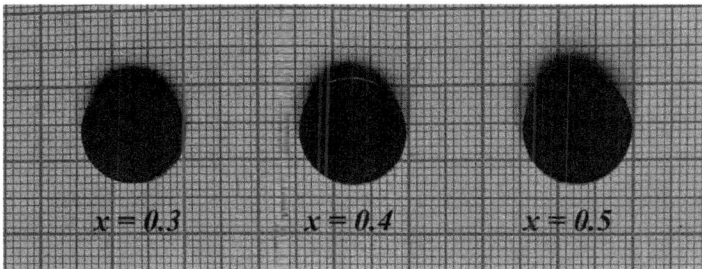

Figure 1.17 *The synthesized samples of $La_{1-x}Sr_xMnO_3$, x=0.3, 0.4 and 0.5.*

1.6   Characterization methods

1.6.1   Powder X-ray diffraction

To understand the structure and properties of a material, one should know how the atoms are arranged in the crystal structures. The spatial arrangement of atoms in the material can be studied through diffraction experiments. In an experiment of diffraction, the incident waves strike on the material and a detector records the outgoing diffracted waves. The scattered waves with constructive interference produce a diffraction pattern. In X-ray diffraction, monochromatic X-rays are generated, collimated and directed towards the sample. The interaction of the incident X-rays with the sample produces constructive interference, when the Bragg's Law [Bragg, 1913] is satisfied. The Bragg's law [Bragg, 1913] relates the wavelength of incident X-rays to the diffraction angle and the spacing between the atomic planes in a crystalline sample. The equation $n\lambda=2d\sin\theta$, is the Bragg's law for X-ray diffraction [Bragg, 1913]. Then the resultant diffracted X-rays are detected and counted. Crystals with perfect periodicity have sharp diffraction peaks whereas crystals with less periodicity have broadened or distorted diffraction peaks. The structure of amorphous materials can also be studied through diffraction peaks [Cullity, 2011].

An X-ray powder diffractometer consists of four essential parts namely, an X-ray tube, the goniometer, the sample and sample-holder and the detector [Azaroff, 1968]. For powder X-ray diffraction, Cu $K_{\alpha1\alpha2}$ doublet ($\lambda = 1.542$ Å) X-rays are collimated and incident on the sample mounted on the goniometer [Stout et al., 1989]. When the goniometer gradually rotates, the sample is being bombarded with X-rays. Constructive interference occurs when the incident X-rays satisfies the Bragg law [Bragg, 1913] and the corresponding intensity peak is recorded by the detector. The detector then converts this recorded signal into a count rate and gives an output via computer monitor or printer. The schematic diagram of an X-ray diffractometer is shown in figure 1.18.

In the present work, powder XRD data sets of the synthesized powder samples were collected with a Bruker AXS D8 advance model X-ray diffractometer.

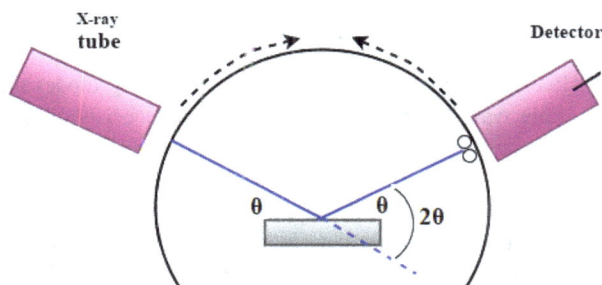

Figure 1.18 Schematic diagram of X-ray diffractometer.

1.6.2 UV-visible spectrophotometry

The UV-visible spectroscopy investigates the electronic transitions of molecules when they absorb energy in the UV (190 nm – 400 nm) and visible (400 nm - 800 nm) regions of the electromagnetic spectrum. It is helpful to study the transmission, absorption and reflectivity of various technologically important materials like coatings, pigments etc. UV-visible spectra have been used for impurity detection, identification of functional groups, structural determination of organic compounds, molecular weight determination and for quantitative measurements.

UV-visible spectrophotometer measures absorbance and transmittance of the sample with respect to the wavelength of the electromagnetic radiation. The major components of a UV-visible spectrophotometer are the light source, monochromator, sample cell and the detector [Gullapalli and Barron, 2010]. The schematic representation of UV-visible spectrophotometer is shown in figure 1.19.

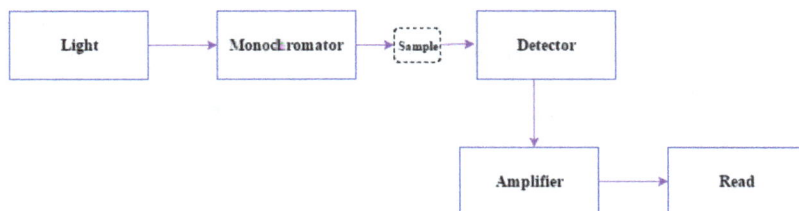

Figure 1.19 Block diagram of UV-visible spectrophotometer.

The radiation sources are a deuterium discharge lamp (190-400nm) for ultra violet (UV) region and a tungsten filament lamp or tungsten halogen lamp for the visible region (300 nm - 2500 nm). Recently, xenon arc lamps and light emitting diodes are used for the ultraviolet and visible wavelengths. The monochromator consists of entrance and exit slits, collimating and focusing lenses and dispersion devices like prisms and holographic gratings. The light radiation from the source is split into two parts, one is the reference and the other passes through sample. The reference beam intensity is taken as zero absorbance. Then the two beams are detected by a detector and the measurement is displayed as the ratio of the two beam intensity values. This ratio of the beam intensities is proportional to the absorbance of the sample. Photo diodes are normally used as detectors. If the instrument has two detectors, then the reference and the sample beam measurements are taken at the same time. The detector converts the light radiation into an electrical signal which is read by a computer.

In the present work, the UV-visible absorption spectra for the samples have been recorded using an UV-visible spectrophotometer Cary 5000 (Varian, Germany).

1.6.3 Scanning electron microscopy

Scanning electron microscope (SEM) is used to analyze the surface morphology of materials. SEM image of the materials is obtained by allowing an electron beam to interact with the sample, which takes place in a vacuum and produces a variety of signals. These signals give information of the surface morphology.

The diagram in figure 1.20 shows the essential parts of a SEM. They are the electron gun, electromagnetic lenses, sample chamber and the detector [Lawes, 1987]. A metallic filament is kept at the top of the SEM set up to generate electrons by thermionic emission and this region is referred to as electron gun. These electrons are then formed into a beam and accelerated down towards the sample. This electron beam is termed as primary electron beam. Then, the electron beam is focused and directed by the electromagnetic lenses so that the beam moves towards the sample. As the electron beam reaches the sample, electrons are knocked out from the surface of the sample due to inelastic collision. These electrons are termed as secondary electrons which are viewed by a detector. Actually, the electron beam scans the sample back and forth, producing an image from the secondary electrons emitted from each spot on the sample.

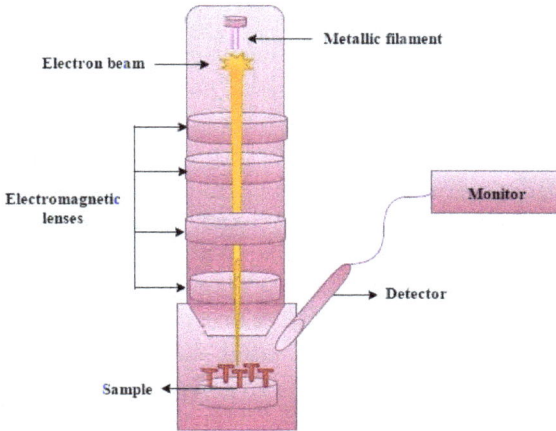

Figure 1.20 Schematic diagram of scanning electron microscope (SEM).

Then, the detector amplifies the signal and sends the signal to the monitor. This entire process takes place inside a vacuum environment. This vacuum is required due to the following three reasons. The first reason is that when the current passes through the filament, it attains high temperature (2700 K). Hence, in the presence of air, metallic filament will oxidize and due to that, it will burn out. Secondly, the operation of electromagnetic lenses needs a clean and dust-free environment. The third reason is that, the dust particles in the air interfere and block the electron beam before it reaches the sample chamber. Hence, a vacuum is needed for the SEM set up.

The scanning electron microscopes used in the present work belong to Carl Zeiss EVO 18 and JSM - 6390LV- JEOL microscopes.

1.6.4 Energy dispersive X-ray spectroscopy

Energy dispersive X-ray spectroscopy (EDS) is a technique used for identifying the elemental composition present in the sample. Generally, the EDS system is equipped with a scanning electron microscope (SEM). The EDS setup consists of four primary components such as the excitation source, sample cell, the X-ray detector and the analyzer [Russ, 1984]. During the EDS analysis, the electron beam interacts with the sample so that it produces a variety of emissions including X-rays. The EDS detector separates the characteristic X-rays of different elements into an energy spectrum. The EDS system software analyzes the energy spectrum to determine the specific elements in the sample and provides elemental composition maps. The block diagram of energy dispersive X-ray spectroscope is given in figure 1.21.

The EDS spectrum is nothing but a plot which gives how frequently an X-ray is absorbed for each energy level of an atom. Each peak in the EDS spectrum represents the energy levels for which the most X-rays had been received. These peaks are unique to an atom and hence denote a single element. The highest peak in the spectrum refers to the more concentrated element in the sample. An EDS spectrum not only identifies the element corresponding to each of its peaks, but the type of X-ray to which it corresponds as well. EDS spectrum is a plot between X-ray counts and energy in keV. Energy peaks correspond to different elements present in the sample. The peaks are narrow and are resolved but some elements have multiple peaks. The X-ray peaks corresponding to elements with low concentration have not been resolved from the background radiation.

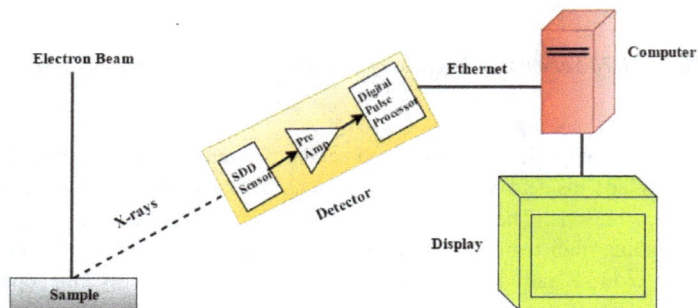

Figure 1.21 Block diagram of energy dispersive X-ray spectroscope (EDS).

In the present work, the elemental compositions of the samples have been analyzed with a Quantax 200 with X-flash-Bruker and JED–2300-JEOL spectroscopes.

1.6.5 Vibrating sample magnetometry

A vibrating sample magnetometer (VSM) is used to measure the magnetic parameters (like saturation magnetization, remanent magnetization, coercive field) of magnetic materials. It operates under the principle of the Faraday's law (Simon Foner, 1959; Wesley Burgei, 2003) of electromagnetic induction

$$\varepsilon = -N\frac{d}{dt}(BA\cos\vartheta) \tag{1.6}$$

Where ε is induced *e.m.f.* in the coil

N is the number of turns in the coil

B is the applied magnetic field

A is the area of the coil

ϑ is angle between applied magnetic field B and the direction normal to the surface of the coil.

The Faraday's law states that a changing magnetic field will produce an electric field. In VSM, the magnetic sample to be studied is placed in a constant magnetic field. This magnetic field will magnetize the sample by aligning the magnetic spins with the field. If the constant magnetic field is stronger, then the magnetization will be larger. Depending upon this magnetic field and magnetization, magnetic materials are classified as ferromagnetic, ferrimagnetic, antiferromagnetic, paramagnetic and diamagnetic materials. These different magnetic behaviors of the materials can be studied through vibrating sample magnetometer (VSM) measurements as a function of magnetic field, time and temperature.

The schematic diagram of VSM is shown in figure 1.22. The essential parts of VSM are an electromagnet with power supply, vibration exciter, sample holder, magnetic sensor coils (pick-up coils), detector (Hall probe), amplifier and computer interface [Simon Foner, 1959]. An electromagnet generates constant magnetic field which is used to magnetize the sample.

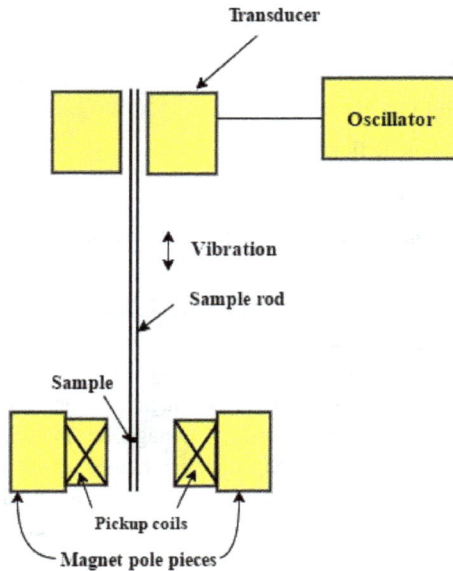

Figure 1.22 *Schematic diagram of vibrating sample magnetometer (VSM).*

The sample holder is attached to the vibration exciter and is hung between the pole pieces of the magnet. The sample moves up and down by means of an exciter. By rotating the sample rod, the desired orientation of the sample is fixed at a constant magnetic field. Now the sample generates an alternating current in these coils with a frequency same as that of the vibration of the sample.

The signal produced gives the details of the magnetization of the sample. This signal created by the sensor coils is amplified by an amplifier. The lock-in amplifier is tuned to pick up only signals at the vibration frequency. The computer interface is used for the data collection and the data can be graphed and plotted.

In this work, the magnetic properties have been analyzed using the Lakeshore VSM 7410 model.

1.7 Methodologies used for analysis

1.7.1 Structural determination using the Rietveld refinement technique

To understand the properties of materials, it is essential to know its atomic structure. The X-ray or neutron diffraction methods are the best techniques to study the atomic structure of materials. There are two diffraction processes, viz;

1. Single-crystal X-ray diffraction

2. Powder X-ray diffraction

The atomic structure of relatively large crystal can be determined by the intensity data obtained from the single crystal X-ray diffraction. But most materials are grow in the form of tiny crystallites (powder). In reality, both the single-crystal and powder X-ray diffraction techniques have their own strengths and weaknesses and one cannot supersede the other. In the past three decades, powder diffraction has played a vital role in the field of structural physics, chemistry and material science. High temperature superconductors and high pressure research have relied mostly on power diffraction techniques. In early 1990s, powder diffraction data have been used for the structural determination of only very few crystals. But today, numerous organic and inorganic crystal structures have been solved through powder diffraction data. Many powder diffraction methodologies have been developed which have contributed to analyze the structure of the compounds. Since the powder diffraction peaks grossly overlap in the conventional powder diffraction method, it prevents the exact determination of the crystal structure. The Rietveld method [Rietveld, 1969] minimizes the impact of these overlapping peaks and determines the real crystal structure. The Rietveld method [Rietveld, 1969] is a technique for the crystal structure refinement which uses the whole powder diffraction pattern, devised by Hugo Rietveld. In the Rietveld method [Rietveld, 1969], the least-squares approach is used so that the calculated and measured diffraction profiles are optimized and with the iterative technique, the profiles are refined. The following sections explain the powder profile parameters that we come across in the Rietveld [Rietveld, 1969] refinement technique.

1.7.1.1 Peak shape

In powder XRD, the shape of a powder diffraction reflection is influenced by the characteristics of the X-ray beam, the sample size and shape and the experimental arrangements. For monochromatic neutron sources, the convolution of the various effects has been found to result in a reflection almost exactly Gaussian in shape. If this distribution is assumed, then the contribution of a given reflection to the profile y_i at position $2\theta_i$ is

$$y_i = I_k \exp[-4 \ln \left(\frac{2}{H_k^2}\right) (2\theta_i - 2\theta_k)^2] \tag{1.7}$$

where H_k is the full width at half-maximum, $2\theta_k$ is the centre of the reflections and I_k is the calculated intensity of the reflections (determined from the structure factor, the Lorentz factor, and multiplicity of the reflection).

Due to the vertical divergence of the beam, at very low diffraction angles, the reflections may acquire an asymmetry. Rietveld [Rietveld, 1969] used a semi-empirical correction factor, A_s to account for this asymmetry

$$A_s = 1 - [\frac{sP(2\theta_i - 2\theta_k)^2}{\tan\theta_k}] \tag{1.8}$$

where P is the asymmetry factor and s is +1, 0, -1 depending on the difference $2\theta_i$-$2\theta_k$ being positive, zero or negative respectively. At a given position, more than one diffraction peak may contribute to the profile. The intensity is simply the sum of all reflections contributing at the point $2\theta_i$.

1.7.1.2 Peak width

At higher Bragg angle side of the XRD pattern, the widths of the diffraction peaks are found to be broadened. This angular dependency was originally represented by

$$H_k^2 = U\tan^2\theta_k + V\tan\theta_k + W \tag{1.9}$$

where U, V and W are the half width parameters and may be refined during the fit.

1.7.1.3 Preferred orientation

In powder samples, the crystallites will be like a plate or rod. These crystallites have a tendency to align themselves along the axis of a cylindrical sample holder. In solid polycrystalline samples, the crystallization of the material may result in greater volume fraction of certain crystal orientations (usually referred to as texture). In such cases, the reflection intensities will vary from that predicted for a completely random distribution. Rietveld [Rietveld, 1969] allowed for moderate cases of the former by introducing a correction factor:

$$I_{corr} = I_{obs} \exp(-G\alpha^2) \tag{1.10}$$

Where I_{obs} is the intensity expected for a random sample, G is the preferred orientation parameter and α is the acute angle between the scattering vector and the normal of the crystallites.

1.7.1.4 Background function

The background function, y_{bi}, at step i is approximated by a finite sum of Legendre polynomials, $F_j(x_i)$ [Abramowitz and Stegun, 1966], orthogonal relative to integration over the interval [-1, 1]:

$$y_{bi} = \sum_{j=0}^{\Pi} b_j F_j(x_i)$$ (1.11)

$F_j(x_i)$'s for $j \geq 2$ are calculated from $F_{j-1}(x_i)$ and $F_{j-2}(x_i)$ using the relation

$$F_j(x_i) = \left(\frac{2j \pm 1}{j}\right) x_i \ F_{j\pm1}(x_i) \pm \left(\frac{j \pm 1}{j}\right) F_{j\pm2}(x_i)$$ (1.12)

with $F_0(x_i) = 1$ and $F_1(x_i) = x_i$. The coefficients, b_j, are background parameters to be refined in Rietveld analysis, and the variable, x_i, is the diffraction angle, $2\theta_i$, normalized between -1 and 1

$$x_i = \frac{2\theta_i - \theta_{max} - \theta_{min}}{\theta_{max} - \theta_{min}}$$ (1.13)

Using this background function, correlation coefficients between background parameters can be somewhat reduced. The "humps" due to amorphous or poorly crystallized compounds may be fitted well by increasing the number of refinable background parameters. However, care must be taken for not to vary too many background parameters when dealing with a diffraction pattern whose background has simple dependence on 2θ.

1.7.1.5 Refinement procedure

The principle behind the Rietveld refinement [Rietveld, 1969] technique is to minimize a function M which analyzes the difference between a calculated profile y(calc) and the observed data y(obs). Rietveld [Rietveld, 1969] defined such a function M as:

$$M = \sum_i W_i \left\{y_i^{obs} - \frac{1}{c}y_i^{calc}\right\}^2$$ (1.14)

Where W_i is the statistical weight and c is an overall scale factor such that $y_i^{calc} = cy_i^{obs}$.

The approach behind the Rietveld [Rietveld, 1969] technique is to calculate the entire powder pattern using various refinable parameters and by selecting these parameters appropriately to minimize the weighted sum of the squared differences between the observed and the calculated powder XRD pattern using the least squares methods. In that way, the intrinsic problem of the powder diffraction method with systematic and accidental peak overlap is overcome. At present, the Rietveld method [Rietveld, 1969] has been so successful in determining the structure of materials which are in the form of powders. Moreover, this method is also applied in determining the components of the chemical mixtures.

1.7.1.6 Rietveld refinement with JANA 2006

The Rietveld refinement technique was carried out using the software JANA 2000 [Petříček et al., 2000] and its improvised version JANA 2006 [Petříček et al., 2006]. In the present work, the powder XRD profile refinement with the help of the Rietveld [Rietveld, 1969] technique was carried out using JANA 2006 [Petříček et al., 2006]. The refinement software JANA 2006 [Petříček et al., 2006] is useful for the analysis of the crystal structures and composite crystals as well as aperiodic crystals.

In the Rietveld refinement [Rietveld, 1969] technique using the JANA 2006 [Petříček et al., 2006] software, the observed XRD profiles are matched with the profiles constructed similarly by using Gaussian FWHM parameters, Scherer coefficient for Gaussian broadening [Thompson, 1987], pseudo-voigt [Wertheim, 1974] profile shape function of Thompson and symmetric profile shape function [Howard, 1982]. JANA 2006 [Petříček et al., 2006] also employs the correction for preferred orientation using March - Dollase function [March, 1932; Dollase, 1986]. The calculated profiles thus evolved are compared with the observed ones. Finally, the structure factors evolved from the Rietveld refinements [Rietveld, 1969] were further utilized for the estimation of charge density in the unit cell.

1.7.2 Electron density distribution

The quantum mechanical theory explains that the electron density is the measure of the probability of an electron being present at a specific location. The atoms are surrounded by electron clouds. The electron density is defined as the number of electrons per unit volume. The chemical bondings as well as the physical and chemical properties of the crystal systems have been analyzed by the electron density of a system. The electron density distribution study is applied in many disciplines in chemistry, physics, biology and in geology [Stout, 1970]. The study of chemical bonding and internal local structure of a crystalline system is very important and it gives useful information about the transport properties which can be effectively utilized for device applications. For the precise understanding of the nature of chemical bonds, it is essential to study the electron density distribution between the atoms.

The electron density is obtained from the structure factor as

$$\rho(r) = \frac{1}{V}\sum_H F(H)\exp(-2\pi i \mathbf{H}.\mathbf{r}) \qquad (1.15)$$

Where V is the unit cell volume, F(H) are structure factors and H are indices denoting a particular scattering direction corresponding to a crystal plane.

Since crystal lattice is arranged in a periodic manner, the electron density in crystals is also considered to behave as a periodic function. The unit cell of the crystal lattice is divided into small volumes dV in which there are $\rho(r)dV$ number of electrons. In X-ray scattering experiment, the wavelet scattered by this volume element dV is,

$$\rho (x,y,z) \exp[-2\pi i(hx + ky + lz)]dV \tag{1.16}$$

For all the elements in the unit cell, the resultant sum of contributions i.e., the integral over its volume gives,

$$F_{hkl} = \int \rho(x,y,z) \exp[-2\pi i(hx - ky + lz)]\, dV \tag{1.17}$$

Hence, the structure factor is considered as a resultant of adding the scattered waves in the direction of the *hkl* reflection from the atoms in the unit cell. This approach was based on the assumption that the scattering power of the electron cloud surrounding each atom could be equated to that of the proper number of electrons concentrated at the atomic centre.

By knowing the structure factors and phases, the electron density distribution of the unit cell can be calculated. There is necessarily a one to one relationship between structure magnitudes and electron density, *i.e* given set of magnitudes must correspond to one and only one electron density distribution. The magnitudes of individual structure factors are calculated as the square root of the measured diffraction intensity and their phases are determined by solving the structure The interpretation is described as a model, which is improved by the least-squares refinement based on the structure factors. The electron density can then be calculated as a Fourier summation of phased structure factors.

The intensities of diffracted X-rays are due to interference effects of X-rays scattered by all the different atoms in the structure. The diffraction pattern is the Fourier transform of the crystal structure, corresponding to the pattern of waves scattered from an incident X-ray beam by a single crystal; it can be measured by experiment because the amplitudes are obtainable from the directly measured intensities via a number of correction, but the relative phases of the scattered waves are lost, and it can be calculated for a known structure. So, the crystal structure is the Fourier transform of the diffraction pattern and is expressed in terms of electron density distribution concentrated in atoms; it cannot be measured by direct experiment, because the scattered X-rays cannot be refracted by lenses to form an image as done with light in an optical microscope, and it cannot be obtained directly by calculation, because the required relative phases of the waves are unknown. Hence, if it is given a set of structure factors, and we can calculate the electron density distribution, using the Fourier series.

1.7.3 Charge density derived from Fourier method

Crystals are described by periodic functions since they have periodic structures. Of these, a series of cosine and sine terms with appropriate coefficients and with arguments that are successive multiples of x have proven to be most useful. Such series are termed Fourier series. The theoretical implication of Fourier series has wide application in crystallography.

It is assumed now that the three dimensional periodic electron density in a crystal can be represented by a three dimensional Fourier series.

$$\rho(x,y,z) = \sum_{h'}\sum_{k'}\sum_{l'} C_{h'k'l'}\, e^{2\pi i(h'x+k'y+l'z)} \tag{1.18}$$

Where h', k' and l' are integers between $-\infty$ and ∞.

In order to determine the coefficients $C_{h'k'l'}$ in the three dimensional Fourier series representing electron density from equation 1.18, we have,

$$F_{hkl} = \int_V \sum_{h'}\sum_{k'}\sum_{l'} C_{h'k'l'}\, e^{2\pi i(h'x+k'y+l'z)} e^{2\pi i(hx+ky+lz)} \mathrm{d}V \tag{1.19}$$

$$F_{hkl} = \int_V \sum_{h'}\sum_{k'}\sum_{l'} C_{h'k'l'} e^{2\pi i[(h+h')x+(k+k')y+(l+l')z]} \mathrm{d}V \tag{1.20}$$

The exponential is periodic, and the integral over one period is zero for all terms except that one for which $h' = -h,\ k' = -k, l' = -l$. In this case, the periodicity disappears and

$$F_{hkl} = \int_V C_{\overline{hkl}}\, \mathrm{d}V = V C_{\overline{hkl}} \tag{1.21}$$

$$C_{\overline{hkl}} = \left(\frac{1}{V}\right) F_{hkl} \tag{1.22}$$

In equation 1.18, substituting $\overline{h}, \overline{k}, \overline{l}$ for h', k', l' and $\left(\frac{1}{V}\right) F_{hkl}$ for $C_{\overline{hkl}}$, and remembering that summing over $\overline{h}, \overline{k}, \overline{l}$ carries the same meaning as summing over h, k, l, lead to the series

$$\rho(x,y,z) = \frac{1}{V}\sum_{h'}\sum_{k'}\sum_{l'} F_{hkl} e^{-2\pi i(hx+ky+lz)} \tag{1.23}$$

Equation 1.18 is an expression for the electron density in direct space in terms of the structure factors in reciprocal space, the electron density is the Fourier transform of the structure factors, while the structure factors are in turn the Fourier transform of the electron density. Equation 1.18 can be written as

$$\rho(r) = \frac{1}{V} \sum F_H \, e^{-2\pi i (\overline{H}.\vec{r})} \tag{1.24}$$

This notation is more compact and convenient for general discussions of the Fourier transforms. The experimental determination of charge density using conventional Fourier analysis of observed structure factors has its own limit due to the unrealistic negative charge densities and so a perfect understanding of the charge distribution in solids needs to have better method.

1.7.2.2 Maximum entropy method

To understand the physical and chemical properties of molecular systems, one should require the knowledge of their charge distribution. The maximum entropy method (MEM) [Collins, 1982] has been developed for charge density reconstruction. Using a limited number of diffraction data itself, MEM [Collins, 1982] can give a high resolution density distribution. The resultant density distribution gives detailed information of the the structure, without using a structural model. The MEM [Collins, 1982] electron density map is an accurate mathematical tool for structural analysis. Compared to the map drawn by conventional Fourier transformation, the resolution of the MEM electron density map is higher [Sakata and Sato, 1990]. MEM uses the structure factors retrieved from the Rietveld [Rietveld, 1969] refinement. Hence, the combination of the MEM and the Rietveld [Rietveld, 1969] method provides a detailed structure model. Using XRD, it is impossible to collect the exact values of all the structure factors. MEM [Collins, 1982] introduces the concept of entropy to tackle the uncertainty properly. So, the concept behind the maximum entropy method (MEM) [Collins, 1982] is to obtain the electron density distribution which is consistent with the observed structure factors and to leave the uncertainties to a minimum. The mathematical description of MEM [Collins, 1982] is explained below.

The maximum entropy method is an information-theory-based technique to enhance the information obtained from noisy data [Gull and Daniel, 1978]. The theory is based on the same equations that are the foundation of statistical thermodynamics. Both the statistical entropy and the information entropy deal with the most probable distribution. In the case of statistical thermodynamics, this is the distribution of the particles over position and momentum space ("phase space"), while in the case of information theory, the distribution of numerical quantities over the ensemble of pixels is considered. The probability of a distribution of N identical particles over m boxes, each populated by n_i particles, is given by

$$P = \frac{N!}{n_1! n_2! n_3! ... n_m!}$$

(1.25)

As in statistical thermodynamics, the entropy is defined as $ln(P)$. Since the numerator is constant, the entropy is, apart from a constant, equal to

$$S = -\sum_i n_i \ln n_i$$

(1.26)

Where, Stirlings' formula $\ln N! \approx N \ln N - N$ has been used.

In case there is a prior probability q_i for box i to contain n_i particles, expression (eq. 1.25) becomes

$$P = \frac{N!}{n_1! n_2! n_3! ... n_m!} q_1^{n_1} q_2^{n_2} ... q_m^{n_m}$$

(1.27)

which gives, for the entropy expression,

$$S = -\sum_i n_i \ln n_i + \sum_i n_i \ln q_i = -\sum_{i=1}^{m} n_i \ln \frac{n_i}{q_i}$$

(1.28)

The maximum entropy method was first introduced into crystallography by Collins (1982), who, based on equation 1.28, expressed the information entropy of the electron density distribution as a sum over M grid points in the unit cell, using the entropy formula (Jaynes, 1968)

$$S = -\sum \rho'(r) \ln\left(\frac{\rho'(r)}{\tau'(r)}\right)$$

(1.29)

Where, the probability $\rho'(r)$ and prior probability $\tau'(r)$ are related to the actual electron density in a unit cell as,

$$\rho'(r) = \frac{\rho(r)}{\sum_r \rho(r)} \quad \text{and} \quad \tau'(r) = \frac{\tau(r)}{\sum_r \tau(r)}$$

(1.30)

Where, $\rho(r)$ is the electron density and $\tau(r)$ is the prior electron density at a certain fixed r in a unit cell. Hereafter, in this theory, the actual densities are treated instead of normalized densities. And $\rho'(r)$ becomes $\tau'(r)$ when there is no information. The $\rho'(r)$ and $\tau'(r)$ are normalized as

$$\sum \rho'(r) = 1 \quad \text{and} \quad \sum \tau'(r) = 1$$

(1.31)

A constraint C is introduced here as,

$$C = \frac{1}{N}\sum_k \frac{|F_{CAL}(k) - F_{OBS}(k)|^2}{\sigma^2(k)}$$

(1.32)

Where N is the number of reflections used for MEM analysis, $\sigma(k)$ is the standard deviation of $F_{OBS}(k)$, the observed structure factor and $F_{CAL}(k)$ is the calculated structure factor given by

$$F_{cal}(k) = V \sum \rho(r) \exp(-2\pi i \mathbf{k}.\mathbf{r})\, dV$$

(1.33)

The constraint C is sometimes termed as a weak constraint, in which the calculated structure factors agree with the observed structure factors as a whole when the constraint C becomes unity. Equation (1.33) indicates, the structure factors are given by the Fourier transform of the electron density distribution in a unit cell. In the MEM analysis [Collins, 1982], there is no need to introduce the atomic form factors, by which the structure factors are normally written. Equation 1.33 guarantees that it is possible to introduce any type of deformation of the electron densities ρ(r), in real space as long as information concerning such a deformation is included in the observed data.

Here, while maximizing the entropy, Lagrange's method of undetermined multiplier (λ) is used in order to constrain the function C to be unity. Then, we have,

$$Q = S - \left(\frac{\lambda}{2}\right)C \quad = -\sum \rho'(r)\ln\left(\frac{\rho'(r)}{\tau'(r)}\right) - \frac{\lambda}{2N}\sum_k \frac{|F_{CAL}(k) - F_{OBS}(k)|^2}{\sigma^2(k)}$$

(1.34)

and when $\frac{dQ}{d\rho} = 0$ and using the approximation,

ln x = x-1 we get,

$$\rho(\mathbf{r}_i) = \tau(\mathbf{r}_i)\exp\left\{\left(\frac{\lambda F_{000}}{N}\right)\left[\sum \frac{1}{\sigma(k)^2}\right]|F_{cts}(\mathbf{k})\text{-}F_{cal}(\mathbf{k})|\exp(-2\pi j\, \mathbf{k}.\mathbf{r})\right\}$$

(1.35)

Where, $F_{000}=Z$, the total number of electrons in a unit cell. To solve equation 1.35 in a simple way, the following approximation has been introduced which replaces $F_{cal}(\mathbf{k})$ as,

$$F_{cal}(k) = V \sum \tau(r)\exp(-2\pi i \mathbf{k}.\mathbf{r})dV$$

(1.36)

This approximation can be named zero[th] order single pixel approximation (ZSPA). Hence, using this approximation, the right hand side of eq. 1.35 becomes independent of $\tau(\mathbf{r})$ and eq. 1.35 can be solved in an iterative manner starting from a given initial density for the prior distribution. A uniform density distribution is given as the prior density $\tau(r)$ as $0 \leq \tau(r) \geq Z/_M$ where, M is the number of pixels for which the electron density is calculated. The reason for this choice of prior distribution is that the uniform density distribution corresponds to the maximum entropy state among all possible density distributions. The validity of ZSPA has been fully discussed based on the simple two pixel model which can be analytically solved. In the calculation of $\rho(r)$, all of the symmetry recruitments are satisfied and the number of electrons (Z) is always kept constant through an iteration process. Mathematically, the summation concerning $\rho(r)$ in the above equations should be written as an integral. Since we must use a very limited number of pixels in the numerical calculation, the integral is replaced by the summation in the above equations.

After completion of the MEM enhancement, it becomes possible to evaluate the reflections missing from the summation. In a Fourier summation, the amplitudes of the unobserved reflections are assumed to be equal to zero, while the MEM technique [Collins, 1982] provides the most probable values.

1.7.2.3 Methodology for the determination of charge density

The present book focuses mainly on MEM analysis [Collins, 1982] and the Rietveld [Rietveld, 1969] refinement method. If the structure factors are refined using the Rietveld refinement [Rietveld, 1969], then they are further utilized for the evaluation of MEM charge density. In the MEM refinement process, the analysis was performed for all data sets using the Fortran 90 program PRIMA [Ruben and Fujio, 2004], to get a 3D density file. The input file contains the cell parameters, pixels, space group, Lagrange parameter, total charge and structure factors. During the refinement, the uniform prior density and Lagrangian multiple are applied to get the preference file. The refinement process continues until the constraint C reaches 1. Finally, with this density file, the 3D electron density iso-surface has been visualized using the VESTA [Momma and Izumi, 2006] software package. To understand the nature of the bond in the material, two-dimensional (2D) and one dimensional (1D) distribution of electron density on different lattice planes are discussed in the following chapter.

1.7.3 Optical band gap determination

The band gap energy is the minimum amount of energy required for an electron to undergo a transition from valence band to conduction band. In this work, the UV-visible absorption spectra of the samples have been used for the determination of the energy band gap. The absorbance (α), the photon energy (hv) and the optical band gap (E_g) are related by Wood and Tauc relation [Wood and Tauc, 1972] as,

$$\alpha h v = A(h v - E_g)^n \tag{1.37}$$

where, A is an energy independent constant, sometimes named as band tailing parameter. The constant "n" in equation (1.37) depends on the nature of the material (crystalline or amorphous) and the photon transition. For the direct allowed transitions $n = 1/2$ and for indirect allowed transitions $n = 2$. For the direct band gap materials, $n = 1/2$. Hence, equation (1.37) becomes,

$$(\alpha h v)^2 = \alpha_0 (h v - E_g) \tag{1.38}$$

Equation (1.38) resembles the equation of a straight line,

$$y = mx + c \tag{1.39}$$

Comparison of equations (1.38) and (1.39) gives,

$y = (\alpha h v)^2$; If $y = 0$, then, $\alpha_0 (h v - E_g) = 0$;

But $\alpha_0 \neq 0$; Therefore $(h v - E_g) = 0$;

Hence, $E_g = h v$

Hence, to determine the optical band gap of the materials, using the UV-visible absorption data, Tauc plot between $(\alpha h v)^2$ vs (hv) has been drawn in this work. By extrapolating the linear portion of the Tauc plot to meet the x-axis at $(\alpha h v)^2 = 0$, the value of the energy band gap can be determined.

1.7.4 Grain size determination

The average grain size of the samples was estimated from the powder X-ray diffraction data using the Scherrer formula [Cullity, 2001] through GRAIN software [Saravanan , 2008]. The Scherrer formula [Cullity, 2001] is given as,

$$t = \frac{0.9\,\lambda}{\beta\,cos\theta} \qquad \cdots\cdots \tag{1.40}$$

     where t is grain size (size of the coherently diffracting domain)

     λ is wavelength of X-ray used

     β is the full width at half maximum

     θ is the Bragg angle

The input data for the GRAIN software [Saravanan, 2008] were the Bragg angles and their corresponding full-width at half maxima (FWHM) from the powder X-ray data sets.

The output data obtained from the software gives the average grain size of the corresponding samples.

References

[1]   Abdel-Latif I. A., J. Phys, 1, 15 (2012)

[2]   Abramowitz M and Stegun I. A, 'Handbook of Mathematical Functions', National Bureau of Standards, (1964)

[3]   Azaroff L.V., Elements of X-ray crystallography, Mc Graw hill book company, New York, (1968) p.79

[4]   Anderson P.W., Hasegawa H., Phys. Rev., 100, 675 (1955) https://doi.org/10.1103/PhysRev.100.675

[5]   Athawale A.A., Desai P.A., Ceram. Int, 37, 3037 (2011) https://doi.org/10.1016/j.ceramint.2011.05.008

[6]   Balcells L., Enrich R., Mora J., Calleja A., Fontcuberta J. and Obradors X., Appl. Phys. Lett., 69, 1486 (1996) https://doi.org/10.1063/1.116916

[7]   Bella R. J., Millara G. J. and Drennan J., Solid State Ionics, 131, 211 (2000) https://doi.org/10.1016/S0167-2738(00)00668-8

[8]   Bellakki MB, Shivakumara C., Vasanthacharya N.Y., Prakash A.S., Mater. Res. Bull., 45, 1685 (2010) https://doi.org/10.1016/j.materresbull.2010.06.063

[9]   Bhalla A S., Guo R and Roy R., Mat. Res. Innovat. 4, 3 (2000) https://doi.org/10.1007/s100190000062

[10]   Bragg W.L., Proceedings of the Cambridge Philosophical Society, 17, 43 (1913)

[11]   Brichzin V., Fleig J., Habermeier H.U. and Maier J., Electrochem. Solid-State Lett., 3, 403 (2000) https://doi.org/10.1149/1.1391160

[12]   Carlos Va'zquez-Va'zquez, M. Arturo Lo'pez-Quintela, J. Solid State Chem., 179, 3229 (2006)

[13]   Chen X., Hou P.Y., Jacobson C.P., Visco S. J., DeJonghe C., Solid State Ionics 176, 425 (2005) https://doi.org/10.1016/j.ssi.2004.10.004

[14]   Coey J M D, Viret M, von Molnaír S, Adv. Phys., 48, 167 (1999) https://doi.org/10.1080/000187399243455

[15]   Collins D. M., Nature, 298, 49 (1982) https://doi.org/10.1038/298049a0

[16]   Corrêa H. P. S., Paiva-Santos C O., Setz L. F., Martinez L. G., Mello-Castanho S.
       R. H., Orlando M. T. D., Powder Diffraction Suppl., 23, 18 (2008)
       https://doi.org/10.1154/1.2903501

[17]   Cullity B.D., S.R. Stock, Elements of X-ray Diffraction, 3rd ed. (Prentice Hall,
       New Jersy), 167 (2001)

[18]   David W. L., Montgomery F. C., Armstrong T. R., Sens. Actuators B 111-112, 84
       (2005) https://doi.org/10.1016/j.snb.2005.06.043

[19]   Daengsakul S., Mongkolkachit C., Thomas C., Ian thomas, Sineenatsiri,
       Amornkitbamrung V., Maensiri S., Optoelectron. Adv. Mat., 3, 106 (2009)

[20]   Dollase W.A.J., Appl. Crystallogr., 19, 267 (1986)
       https://doi.org/10.1107/S0021889886089458

[21]   Glazer A. M., Acta Cristallogr. A, 31, 756 (1975)
       https://doi.org/10.1107/S0567739475001635

[22]   Glazer A. M., Phase Transitions, 84, 405 (2011)
       https://doi.org/10.1080/01411594.2010.544732

[23]   Goldschmidt V.M., Naturwissenschafsen 14, 477-485 (1926)
       https://doi.org/10.1007/BF01507527

[24]   Gull S.F. and Daniell G.J., Nature, 272, 686 (1978)
       https://doi.org/10.1038/272686a0

[25]   Gullapalli S, Barron A. Characterization of Group 12-16 (II-VI) Semiconductor
       Nanoparticles by UV-visible Spectroscopy, OpenStax CNX Web site.
       http://cnx.org/content/m34601/1.1/, June 12, 2010

[26]   Hayashi S, Sofue S, and Yoshikado S., Electrical Engineering in Japan, 139, 18
       (2002) https://doi.org/10.1002/eej.1156

[27]   Hilpert K., Das D., Miller M., Peck D. H. and Wei R., J. Electrochem. Soc. 143,
       3642 (1996) https://doi.org/10.1149/1.1837264

[28]   Howard C.J., J. Appl. Crystallogr. 15, 615 (1982)
       https://doi.org/10.1107/S0021889882012783

[29]   Ifrah S., Kaddomi A., Gelin P. and Bergeret G., Catal. Commun. 8, 2257 (2007)
       https://doi.org/10.1016/j.catcom.2007.04.039

[30]   Jain M., Li Y., Hundley M.F., Hawley M., Maiorov B., Campbell I.H., Civale,
       L.and Jiab Q.X., Appl. Phys. Lett., 88, 1 (2006)

[31]   Jaynes E.T., IEEE Trans. Syst. Sci. Cybern. SSC-4, 227 (1968)
       https://doi.org/10.1109/TSSC.1968.300117

[32]   Khetre S. M., Jadhav H V, Bamane S. R., Rasayan J. Chem., 2, 174 (2009)

[33]   Larsen P. H., Hendriksen P. V., Mogenson M., J. Therm. Phys., 49, 1263 (1997)

[34]   Lawes, G. Scanning electron microscopy and X-ray microanalysis: Analytical
       chemistry by open learning, John Wiley & sons, (1987)

[35]   Lira-Herna'ndez A. Ivan, De Jesus, F. S., Corte's-Escobedo,y C. A., and Boları'n-
       Miro' A. M., J. Am. Ceram. Soc., 93, 3474 (2010)

[36]   Loa I., Adler P., Grzechnik, A., Syassen K., Schwarz U., Hanfland M., Rozenberg,
       G. Kh., Gorodetsky P. and Pasternak M. P., Phys. Rev. Lett. 87, 1 (2001)
       https://doi.org/10.1103/PhysRevLett.87.125501

[37]   March A., Z. Kristallogr. 81, 285 (1932)

[38]   Marques R.F.C., Jafelicci Jr., M. C., Paiva-Santos O., Varanda L.C., Godoi
       R.H.M., J. Magn. Magn. Mater., 226-230, 812 (2001)
       https://doi.org/10.1016/S0304-8853(00)01403-7

[39]   Masashi Mori, N. M. Sammes, Solid State Ionics, 146, 301 (2002)

[40]   Michael van den Bossche, Steven McIntosh, J. Catal., 255, 313 (2008)
       https://doi.org/10.1016/j.jcat.2008.02.021

[41]   Michael Ziesey, Phil. Trans. R. Soc. Lond. A, 358, 137 (2000)
       https://doi.org/10.1098/rsta.2000.0524

[42]   Mitchell R H, Perovskites: Modern and ancient (Almaz Press, Ontario (C), 2002)

[43]   Momma K and Izumi F, Commission on Crystallogr. Comput IUCr Newslett. 7,
       106 (2006)

[44]   Moulson A.J. and Herbert J.M., Electroceramics: Materials, Properties,
       Applications, 2nd edn, Chapter 4. Ceramic Conductors, (Wiley online Library,
       2003), p.141

[45]   Nadia M, Omari M, Int. J. Nanoelectronics and Materials, 3, 69 (2010)

[46]   Narayan H, Alemu H, Macheli L and Rao G, Nanotechnology, 20, 255601 (2009)
       https://doi.org/10.1088/0957-4484/20/25/255601

[47]   Navrotsky A., Weidner D. J, Geophys. Monogr. Ser., 45, 146 (1989)

[48]   Nikolina L. Petrova, Dimitar S. Todorovsky, Veselinka G. Vasileva, Cent. Eur. J.
       Chem., 3, 263 (2005)

[49]   Nithya V.D., Jacob Immanuel R.. Senthilkumar S. T., Sanjeeviraja C., Perelshtein
        I., Zitoun D., Kalai Selvan R., Mater. Res. Bull., 47, 1861 (2012)
        https://doi.org/10.1016/j.materresbull.2012.04.068

[50]   Petříček V., Dušek M., Palatinus L., JANA 2000, The crystallographic computing
        system Institute of Physics Academy of sciences of the Czech republic, Praha
        (2000).

[51]   Petříček V., Dušek M., Palatinus L., JANA 2006, The crystallographic computing
        system Institute of Physics Academy of sciences of the Czech republic, Praha
        (2006)

[52]   Petrova N. L, Todorovsky D. S, Vasileva V. G., Central European Journal of
        Chemistry, 3, 263 (2005)

[53]   Pissas M. and Papavassiliou G., J. Phys.: Condens. Matter, 16, 6527 (2004)
        https://doi.org/10.1088/0953-8984/16/36/018

[54]   Pradhan S., Roy G.S., Researcher, 5(3), 63 (2013)

[55]   Prado-Gonjal J, Schmidt R, Romero J-J, Avila D, Amador U, Moran E., Inorg.
        Chem., 52, 313 (2013) https://doi.org/10.1021/ic302000j

[56]   Rabeloa A. A., Cardoso de Macedob M., Melob D. M. A., Paskocimasb C. A,
        Martinellib A. E, Nascimento R. M., Mate. Res., 14, 91 (2011)
        https://doi.org/10.1590/S1516-14392011005000018

[57]   Rashid A., Ahmed A., Ahmad S. N., Shaheen S. A. and Manzoor S. J. Magn.
        Magn. Mater., http://dx.doi.org/10.1016/j.jmmm.2013.07.045
        https://doi.org/10.1016/j.jmmm.2013.07.045

[58]   Reddy Channu V. S., Holze R., Walker E. H., New J. Glass. Ceram., 3, 29 (2013)
        https://doi.org/10.4236/njgc 2013.31005

[59]   Rezlescu N., Rezlescu E., Doroftei C., Popa P.D., Ignat M., Digest Journal of
        Nanomaterials and Biostructures., 8, 581 (2013)

[60]   Rietveld H.M., J. Appl. Crystallogr., 2, 65 (1969)
        https://doi.org/10.1107/S0021889869006558

[61]   Rizzuti A., Leonelli C., Proc. Appl. Ceram., 3, 1 (2009)

[62]   Robert S. Roth, Journal of Research of the National Bureau of Standards, 58, 75
        (1957) https://doi.org/10.6028/jres.058.010

[63]   Ruben A. D and F. Izumi, Super-fast Program PRIMA for the Maximum-Entropy Method, Advanced materials Laboratory, National institute for materials science. 1-1 Namiki, Tsukuba, Ibaraki Japan 305 (2004) 0044

[64]   Russ, J. C. Fundamentals of Energy Dispersive X-ray Analysis, Butterworths, London (1984)

[65]   Russo N., Fino D., Sanacco G. and Speechia V., J. catal., 229, 459 (2005) https://doi.org/10.1016/j.jcat.2004.11.025

[66]   Sakata M and Sato M. Acta Cryst. A, 46, 263 (1990) https://doi.org/10.1107/S0108767389012377

[67]   Saravanan R., GRAIN software, Private Communication, (2008)

[68]   Sasaki, S., Prewitt, C.T.& Bass, J.D. Acta Cryst., C43, 1668 (1987) https://doi.org/10.1107/S0108270187090620

[69]   Setz F. G. Luiz and Mello-Castanho R. H. Sonia, Int. J. Appl. Ceram. Technol., 6, 626 (2009) https://doi.org/10.1111/j.1744-7402.2008.02302.x

[70]   Setz, L.F.G. Santacruz I., Leon-Reina L., De la Torre A.G., Ceram. Int. 41, 1177 (2015) https://doi.org/10.1016/j.ceramint.2014.09.046

[71]   Shu Q., Zhang J., Yan B. and Liu J., Mater. Res. Bull., 44, 649 (2009) https://doi.org/10.1016/j.materresbull.2008.06.022

[72]   Simon Foner, Review of Scientific Instruments, 30, (7) (1959) https://doi.org/10.1063/1.1716679

[73]   Singhal S.C., Solid State Ionics, 152-153, 405 (2002) https://doi.org/10.1016/S0167-2738(02)00349-1

[74]   Stout G.H. and Jensen L.H., X-ray structure determination, chapter 1, 2nd edition, Wiley- Interscience publication (1989)

[75]   Suvorov A. and Shevchik A. P., Refract. Ind. Ceram., 45, 196 (2004) https://doi.org/10.1023/B:REFR.0000036729.24986.e3

[76]   Szytuła A, Acta Phys. Pol. A, 118, 303 (2010) https://doi.org/10.12693/APhysPolA.118.303

[77]   Thompson P., Cox D.E., Hastings J.B., J. Appl. Crystallogr. 20, 79 (1987) https://doi.org/10.1107/S0021889887087090

[78]   Tyson T. A., Phys. Rev. B., 53, 985 (1996) https://doi.org/10.1103/PhysRevB.53.13985

[79] Van Aken B. Bas, Auke Meetsma, Tomioka Y., Tokura Y., and Palstra T. M., Phys. Rev. B, 66, 1 (2002)

[80] Wertheim G. K., Butler M.A., West K.W., Buchanan D.N.E., Rev. Sci. Instrum. 45, 1369 (1974) https://doi.org/10.1063/1.1686503

[81] Wesley Burgei, Michael J. Pechan, and Herbert Jaeger, Am. J. Phys. 71, (8) (2003). DOI: 10.1119/1.1572149 https://doi.org/10.1119/1.1572149

[82] West D. L., Montgomery F.C., Armstrong T. R., Sens. Actuators B 106, 758 (2005) https://doi.org/10.1016/j.snb.2004.09.028

[83] Wood D.L., Tauc J., Phys. Rev. B, 5, 3144 (1972) https://doi.org/10.1103/PhysRevB.5.3144

[84] Xifeng Ding, Yingjia Liu, Ling Cao, Lucun Guo, J. Alloys Compd, 425, 318 (2006) https://doi.org/10.1016/j.jallcom.2006.01.030

[85] Zhu Wei-zhong, Mi Y., J. Zhejiang Univ Sci. 5(12) 1471 (2004) https://doi.org/10.1631/jzus.2004.1471

Chapter 2

Results

Abstract

Chapter 2 gives the results obtained from various characterization techniques such as powder XRD, SEM, EDS, UV-vis and VSM for the synthesized lanthanum chromite materials $(La_{0.8}Ca_{0.2})$ $(Cr_{0.9-x}Co_{0.1}Mn_x)O_3$, $(La_{0.8}Ca_{0.2})$ $(Cr_{0.9-x}Co_{0.1}Fe_x)O_3$ & $(La_{0.8}Ca_{0.2})$ $(Cr_{0.9-x}Co_{0.1}Cu_x)O_3$ and manganite materials $La_{1-x}Ca_xMnO_3$ & $La_{1-x}Sr_xMnO_3$. A detailed account of the results of the materials analyzed is given in the subsections.

Section 2.2 gives the raw XRD patterns, the fitted powder XRD profiles and tables for the refined structural parameters for the synthesized chromite and manganite samples. The SEM images, the EDS spectra and tables for elemental compositions have been given in section 2.3. Section 2.4 gives the UV-visible absorption spectra and tables for optical band gap values for the chromite and manganite materials. Magnetization versus magnetic field (M-H) curves and the magnetic parameters have been given in section 2.5. The results from MEM charge density distribution studies have been presented as 3D, 2D and 1D electron density maps in section 2.6.

Keywords

Powder XRD, Structure, Manganite, Chromite, SEM/EDS, UV-Absorption, Magnetic, VSM, Charge Density

Contents

2.1 Introduction

In the present chapter, all the synthesized doped lanthanum chromite and lanthanum manganite samples have been characterized by powder X-ray diffraction for the structural properties. The surface morphology and the elemental composition of the samples have been analyzed using the scanning electron microscopy technique and energy dispersive X-ray spectroscopy. The optical band gap of the prepared doped lanthanum chromites and lanthanum manganites has been determined from the UV-vis absorption spectra. Magnetic properties of the prepared samples have been studied through vibrating sample magnetometry. The charge density distribution between the atoms in the unit cell and the bonding features have been discussed using the maximum entropy method [Collins, 1982]. In this chapter, the results obtained from various characterization techniques of the synthesized lanthanum chromite and lanthanum manganite samples are presented.

2.2 Structural characterization - Powder X-ray diffraction

The powder XRD data sets for all the synthesized lanthanum chromite and lanthanum manganite samples were collected at the Sophisticated Analytical Instrument Facility (SAIF), Cochin University, Cochin, India, using a Bruker AXS D8 advance X-ray diffractometer with CuK$_\alpha$ monochromatic incident beam (λ=1.54056 Å). The XRD data were collected over the 2θ range from 10° to 120° with a step size of 0.02°.

In this section, the observed powder X-ray diffractograms and the profiles refined using the Rietveld method [Rietveld, 1969] are presented for all the synthesized lanthanum chromite and lanthanum manganite samples. The Rietveld refinement [Rietveld, 1969] was carried out using the software JANA 2006 [Petříček et al., 2014]. The refined structural parameters for all the samples are tabulated as given in tables 2.2, 2.3, 2.4, 2.5 and 2.6.

2.2.1 (Co, Mn) doped (La, Ca) based chromites - (La$_{0.8}$Ca$_{0.2}$)(Cr$_{0.9-x}$Co$_{0.1}$Mn$_x$)O$_3$

The synthesized co-doped lanthanum chromite (La$_{0.8}$Ca$_{0.2}$) (Cr$_{0.9-x}$Co$_{0.1}$Mn$_x$)O$_3$ (x=0.03, 0.06, 0.09 and 0.12) samples have been characterized by powder XRD. The observed powder X-ray diffractograms of synthesized (La$_{0.8}$Ca$_{0.2}$)(Cr$_{0.9-x}$Co$_{0.1}$Mn$_x$)O$_3$, (x=0.03, 0.06, 0.09, 0.12) samples are shown in figure 2.1 (a). The enlarged X-ray diffraction patterns corresponding to (121), (220) and (040) planes for all the Mn compositions are (x=0.03, 0.06, 0.09, 0.12) shown in figure 2.1 (b). The shifting of XRD Bragg peaks for various planes of the synthesized samples are listed in table 2.1.

The powder profile refinement for the synthesized $(La_{0.8}Ca_{0.2})(Cr_{0.9-x}Co_{0.1}Mn_x)O_3$, (x=0.03, 0.06, 0.09 and 0.12) samples have been carried out with the Rietveld [Rietveld, 1969] method using the software JANA 2006 [Petříček et al., 2014]. The refined powder XRD profiles of $(La_{0.8}Ca_{0.2})(Cr_{0.9-x}Co_{0.1}Mn_x)O_3$, (x=0.03, 0.06, 0.09 and 0.12) samples are shown in figures 2.2 (a) - (d). The refined structural parameters and the reliability indices for all the Mn compositions (x=0.03, 0.06, 0.09 and 0.12) are given in table 2.2. The orthorhombic unit cell of $(La_{0.8}Ca_{0.2})(Cr_{0.87}Co_{0.1}Mn_{0.03})O_3$ obtained from VESTA [Momma and Izumi, 2008] is shown in figure 2.3.

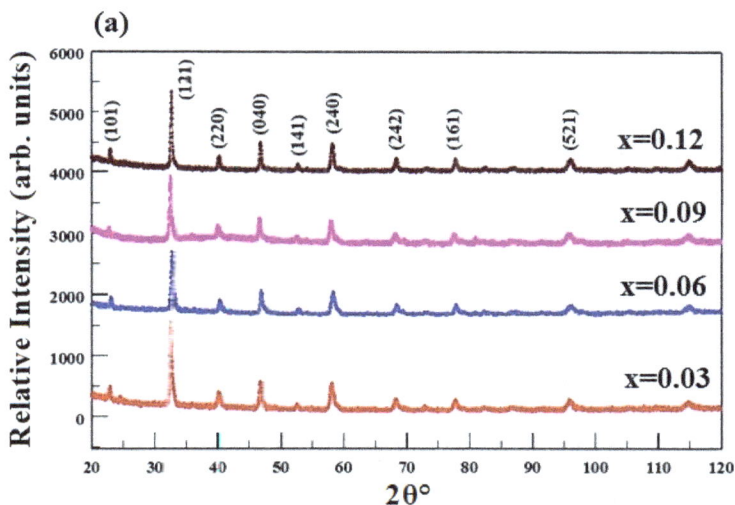

Figure 2.1 (a) Observed X-ray powder diffractograms of $(La_{0.8}Ca_{0.2})(Cr_{0.9-x}Co_{0.1}Mn_x)O_3$, x=0.03, 0.06, 0.09, 0.12.

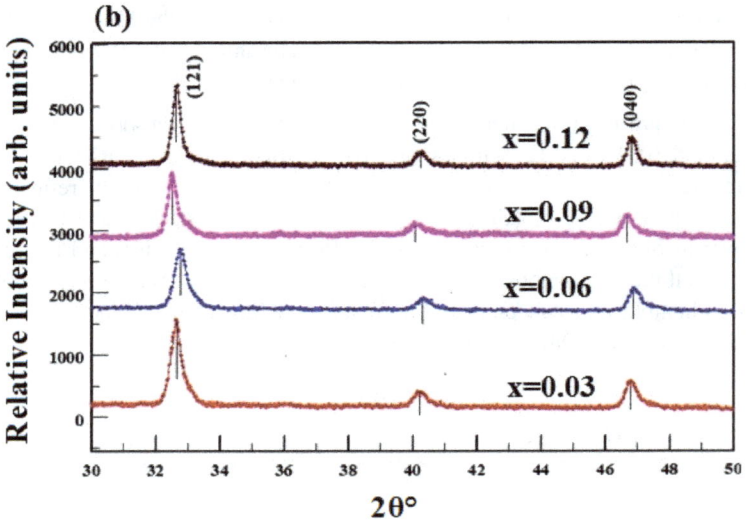

Figure 2.1 (b) *Enlarged XRD peaks of (121), (220) and (040) for $(La_{0.8}Ca_{0.2})(Cr_{0.9-x}Co_{0.1}Mn_x)O_3$, x=0.03, 0.06, 0.09, 0.12.*

Table 2.1 *XRD Bragg peak shifting for $(La_{0.8}Ca_{0.2})(Cr_{0.9-x}Co_{0.1}Mn_x)O_3$, x=0.03, 0.06, 0.09, 0.12.*

(hkl)	Bragg angles (2θ)			
planes	x=0.03	x=0.06	x=0.09	x=0.12
(101)	22.943	23.039	22.810	22.938
(121)	32.648	32.766	32.531	32.656
(220)	40.248	40.368	40.132	40.258
(040)	46.805	46.909	46.705	46.823
(141)	52.531	52.704	52.634	52.725
(240)	58.210	58.311	58.109	58.239
(242)	68.343	68.487	68.301	68.384
(161)	77.850	77.905	77.702	77.829

(a)

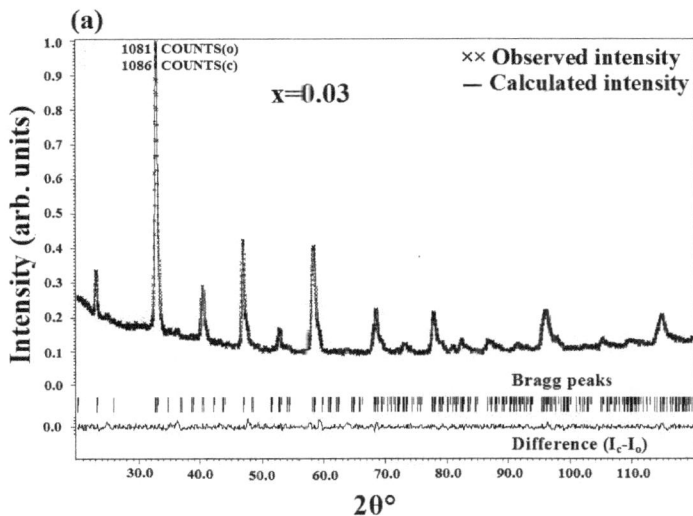

Figure 2.2 (a) Fitted powder XRD profile for $(La_{0.8}Ca_{0.2})(Cr_{0.9-x}Co_{0.1}Mn_x)O_3$, $x=0.03$.

(b)

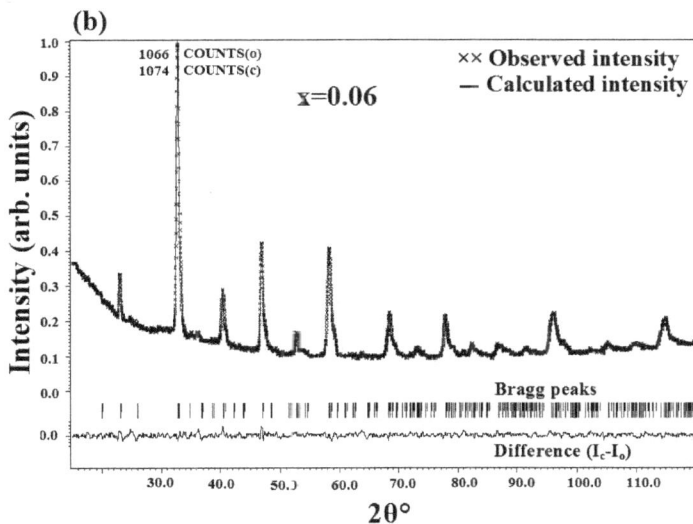

Figure 2.2 (b) Fitted powder XRD profile for $(La_{0.8}Ca_{0.2})(Cr_{0.9-x}Co_{0.1}Mn_x)O_3$, $x=0.06$.

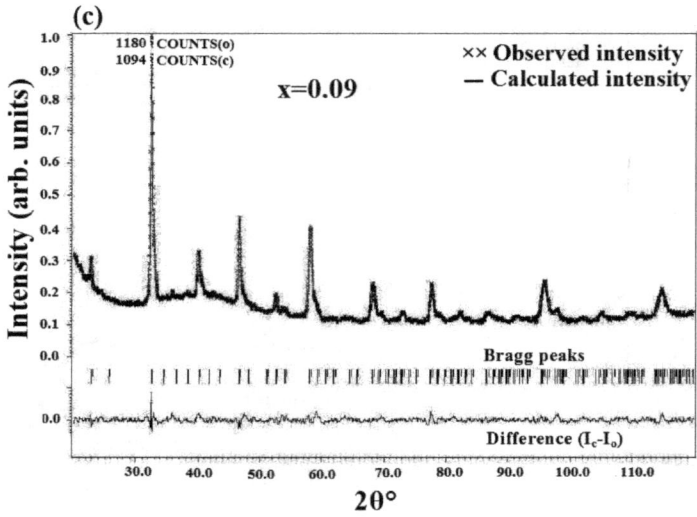

Figure 2.2 (c) *Fitted powder XRD profile for $(La_{0.8}Ca_{0.2})(Cr_{0.9-x}Co_{0.1}Mn_x)O_3$, x=0.09.*

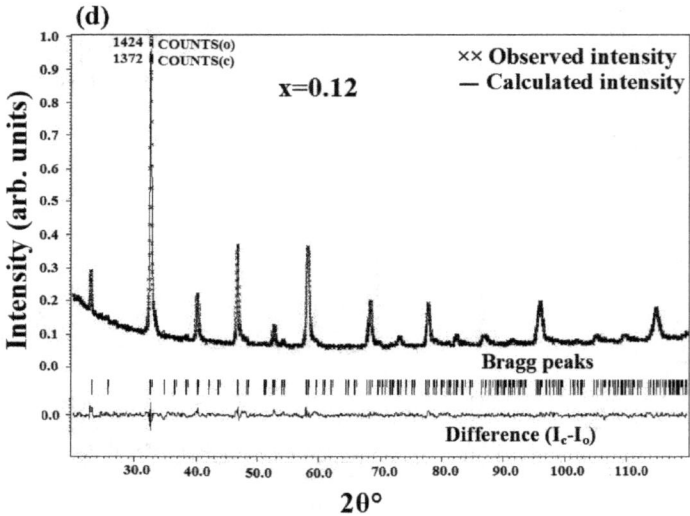

Figure 2.2 (d) *Fitted powder XRD profile for $(La_{0.8}Ca_{0.2})(Cr_{0.9-x}Co_{0.1}Mn_x)O_3$, x=0.12.*

Table 2.2 *Structural parameters for* $(La_{0.8}Ca_{0.2})(Cr_{0.9-x}Co_{0.1}Mn_x)O_3$, $x=0.03$, 0.06, 0.09, 0.12 *through refinement of powder XRD data.*

Parameters	x=0.03	x=0.06	x=0.09	x=0.12
a (Å)	5.474(7)	5.469(9)	5.512(8)	5.474(4)
b (Å)	7.770(11)	7.767(12)	7.799(8)	7.753(6)
c (Å)	5.524(8)	5.522(11)	5.504(8)	5.510(4)
$\alpha=\beta=\gamma$ (°)	90	90	90	90
Unit cell volume (Å³)	235.00(8)	234.60(18)	236.65(6)	233.91(5)
Density (gm/cc)	6.21(22)	6.22(3)	6.17(16)	6.25(12)
R_p (%)	2.88	2.95	2.87	2.83
R_{obs} (%)	2.81	3.65	2.32	3.98
GOF	0.47	0.52	0.51	0.47
$F_{(000)}$	392	392	392	392

R_p- Reliability index for profile
R_{obs} - Reliability index for observed structure factors
GOF - Goodness of fit
$F_{(000)}$ - Number of electrons in the unit cell

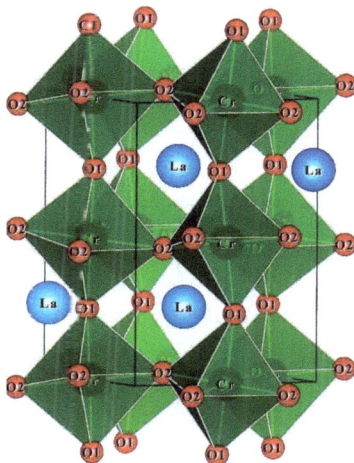

Figure 2.3 *The orthorhombic unit cell of* $(La_{0.8}Ca_{0.2})(Cr_{0.87}Co_{0.1}Mn_{0.03})O_3$.

2.2.2 (Co, Fe) doped (La, Ca) based chromites - $(La_{0.8}Ca_{0.2})(Cr_{0.9-x}Co_{0.1}Fe_x)O_3$

The observed powder X-ray diffractograms of $(La_{0.8}Ca_{0.2})(Cr_{0.9-x}Co_{0.1}Fe_x)O_3$, (x=0.03, 0.06, 0.09, 0.12) samples are shown in figure 2.4 (a). The enlarged XRD peaks corresponding to the (121) plane for all the Fe compositions shown in figure 2.4 (b). The orthorhombic unit cell of $(La_{0.8}Ca_{0.2})(Cr_{0.87}Co_{0.1}Fe_{0.03})O_3$ constructed using VESTA software [Momma and Izumi, 2008] is shown in figure 2.5.

Rietveld refinement [Rietveld, 1969] method has been used to refine the crystal structure model of the synthesized co-doped lanthanum chromite materials utilizing the powder X-ray data sets. Figures 2.6 (a) - (d) show the fitted XRD profiles of prepared chromite samples $(La_{0.8}Ca_{0.2})(Cr_{0.9-x}Co_{0.1}Fe_x)O_3$, (x=0.03, 0.06, 0.09, 0.12) and their refined structural parameters are tabulated in table 2.3.

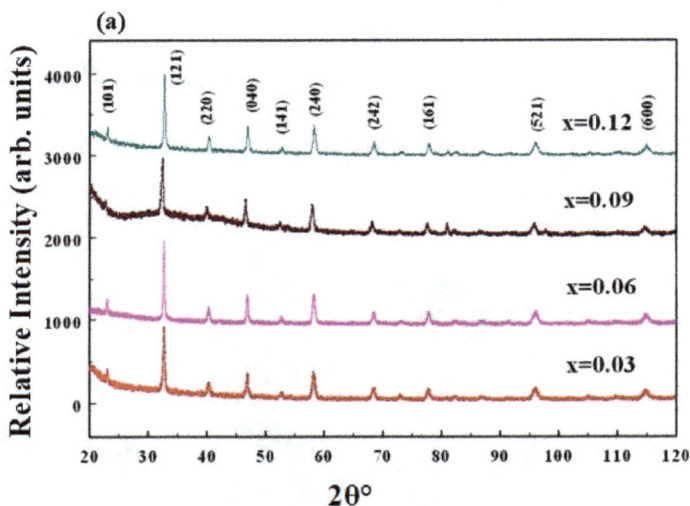

Figure 2.4 (a) Observed X-ray powder diffractograms of $(La_{0.8}Ca_{0.2})(Cr_{0.9-x}Co_{0.1}Fe_x)O_3$, x=0.03, 0.06, 0.09, 0.12.

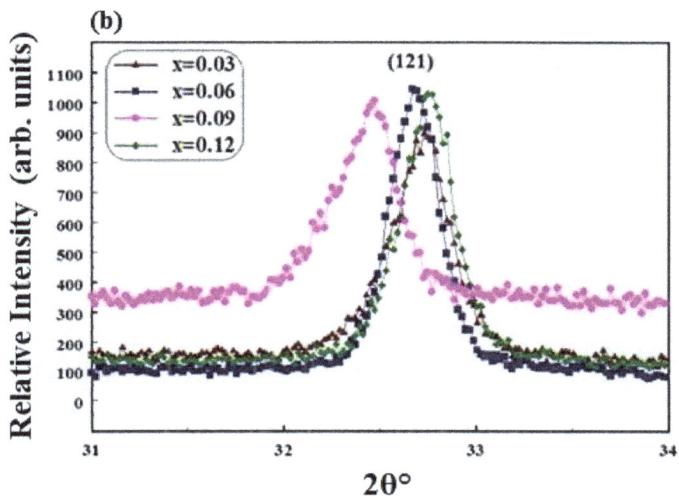

Figure 2.4 (b) *Enlarged XRD peak of (121) for $(La_{0.8}Ca_{0.2})(Cr_{0.9-x}Co_{0.1}Fe_x)O_3$, x=0.03, 0.06, 0.09, 0.12.*

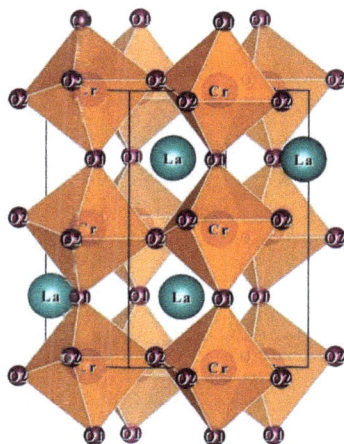

Figure 2.5 *The orthorhombic unit cell of $(La_{0.8}Ca_{0.2})(Cr_{0.87}Co_{0.1}Fe_{0.03})O_3$.*

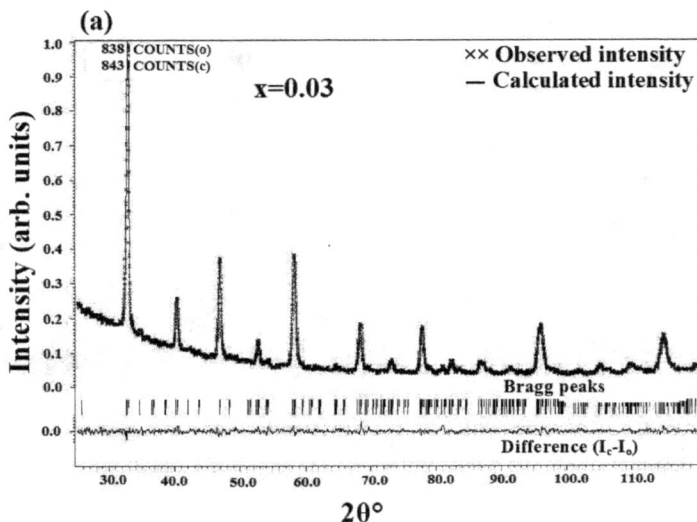

Figure 2.6 (a) Fitted powder XRD profile for $(La_{0.8}Ca_{0.2})(Cr_{0.9-x}Co_{0.1}Fe_x)O_3$, x=0.03.

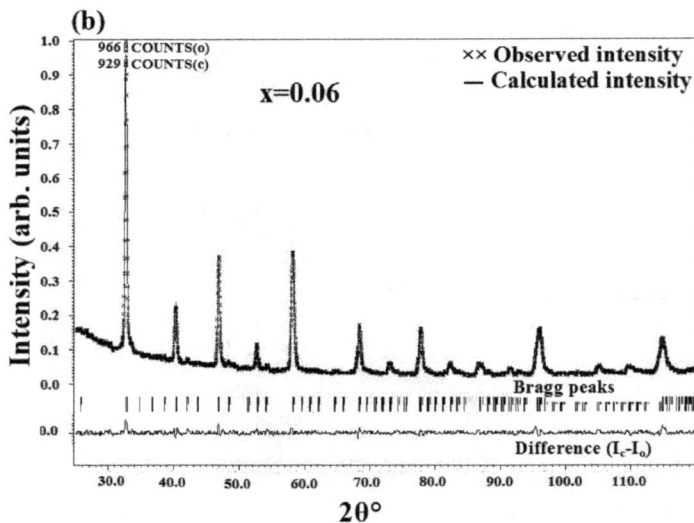

Figure 2.6 (b) Fitted powder XRD profile for $(La_{0.8}Ca_{0.2})(Cr_{0.9-x}Co_{0.1}Fe_x)O_3$, x=0.06.

(c)

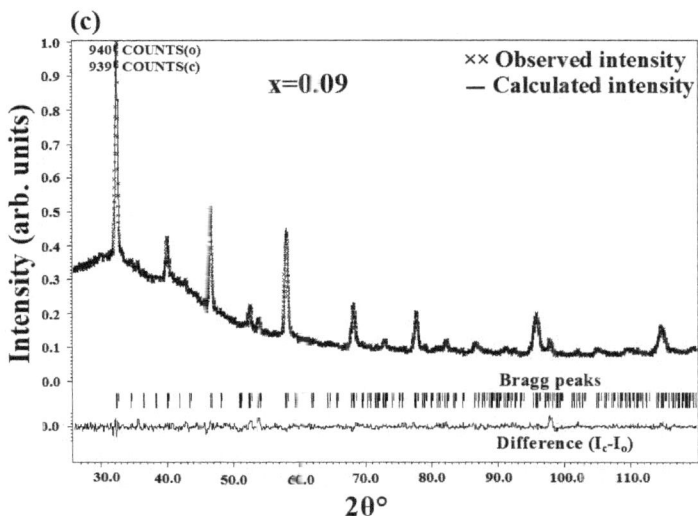

Figure 2.6 (c) Fitted powder XRD profile for $(La_{0.8}Ca_{0.2})(Cr_{0.9-x}Co_{0.1}Fe_x)O_3$, $x=0.09$.

(d)

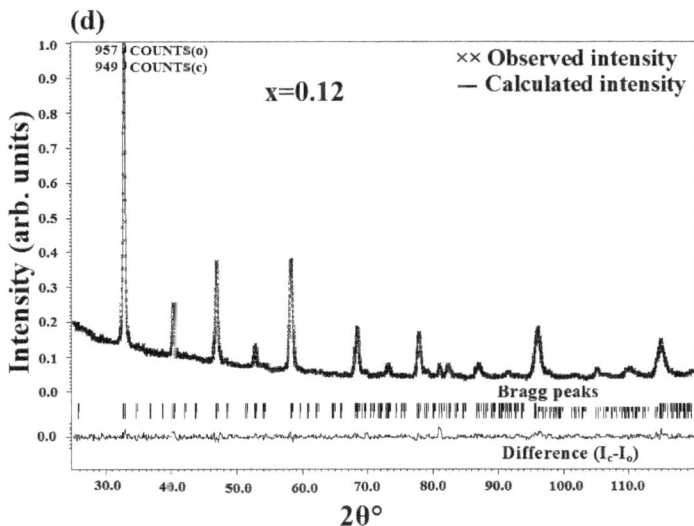

Figure 2.6 (d) Fitted powder XRD profile for $(La_{0.8}Ca_{0.2})(Cr_{0.9-x}Co_{0.1}Fe_x)O_3$, $x=0.12$.

Table 2.3 *Structural parameters for $(La_{0.8}Ca_{0.2})(Cr_{0.9-x}Co_{0.1}Fe_x)O_3$, x=0.03, 0.06, 0.09, 0.12 through refinement of powder XRD data.*

Parameters	x=0.03	x=0.06	x=0.09	x=0.12
a (Å)	5.516(3)	5.507(6)	5.471(5)	5.514(6)
b (Å)	7.772(7)	7.784(5)	7.747(6)	7.757(8)
c (Å)	5.481(3)	5.507(4)	5.506(6)	5.478(6)
$\alpha=\beta=\gamma$ (°)	90	90	90	90
Unit cell volume (Å^3)	235.05(3)	236.13(6)	233.43(5)	234.34(7)
Density (gm/cc)	6.21(9)	6.18(7)	6.26(7)	6.24(9)
R_p (%)	3.60	5.34	2.80	3.35
R_{obs} (%)	3.71	4.57	2.70	3.04
GOF	0.46	0.59	0.60	0.46
$F_{(000)}$	392	392	392	393

R_p- Reliability index for profile
R_{obs}- Reliability index for observed structure factors
GOF- Goodness of fit
$F_{(000)}$ - Number of electrons in the unit cell

2.2.3 (Co, Cu) doped (La, Ca) based chromites - $(La_{0.8}Ca_{0.2})(Cr_{0.9-x}Co_{0.1}Cu_x)O_3$

The observed powder X-ray diffractograms of $(La_{0.8}Ca_{0.2})(Cr_{0.9-x}Co_{0.1}Cu_x)O_3$, (x=0.00, 0.03, 0.12) samples are shown in figure 2.7 (a). The enlarged XRD peaks corresponding to the (121) and (040) planes for all the Cu compositions shown in figures 2.7 (b) and 2.7 (c) respectively.

The orthorhombic unit cell of $(La_{0.8}Ca_{0.2})(Cr_{0.87}Co_{0.1}Cu_{0.03})O_3$ constructed using VESTA [Momma and Izumi, 2008] is shown in figure 2.8. Figures 2.9 (a) - (c) show the fitted Rietveld refined [Rietveld, 1969] plots for the prepared co-doped chromite samples. The refined structural parameters are tabulated in table 2.4.

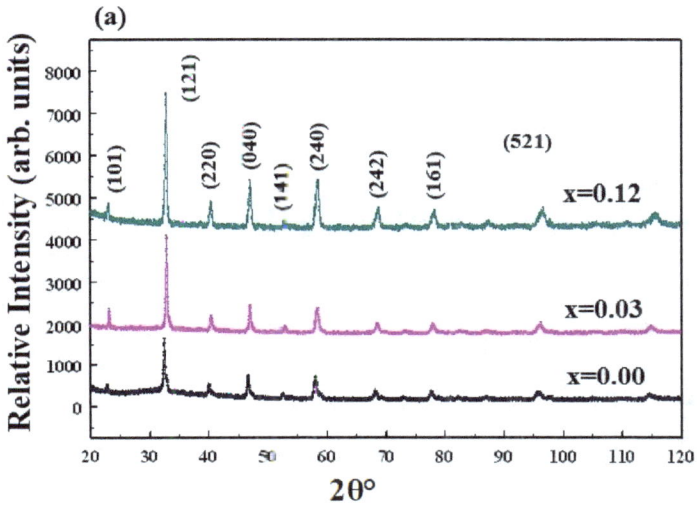

Figure 2.7 (a) *Observed X-ray powder diffractograms of $(La_{0.8}Ca_{0.2})(Cr_{0.9-x}Co_{0.1}Cu_x)O_3$, x=0.00, 0.03, 0.12.*

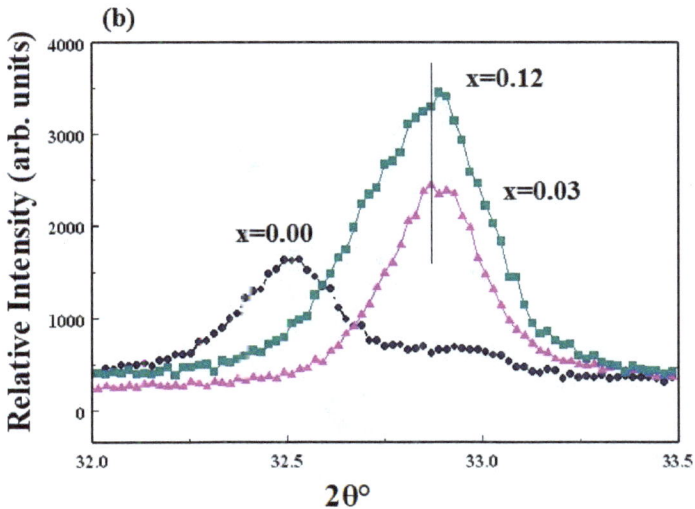

Figure 2.7 (b) *Enlarged XRD peaks of (121) for $(La_{0.8}Ca_{0.2})(Cr_{0.9-x}Co_{0.1}Cu_x)O_3$, x=0.00, 0.03, 0.12.*

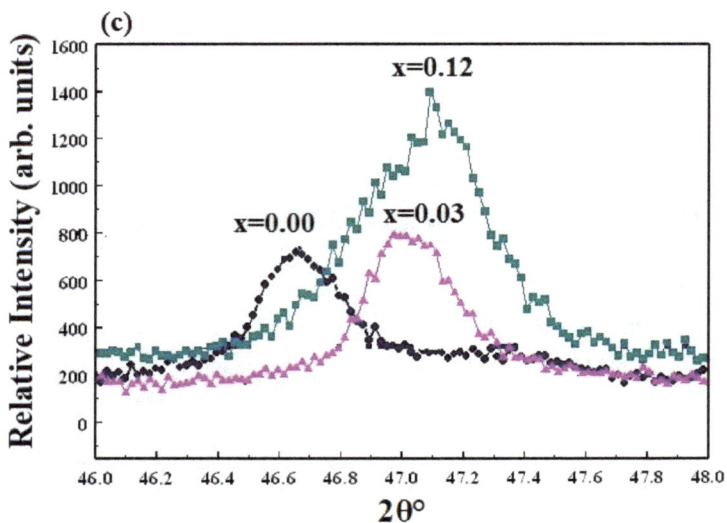

Figure 2.7 (c) *Enlarged XRD peaks of (040) for $(La_{0.8}Ca_{0.2})(Cr_{0.9-x}Co_{0.1}Cu_x)O_3$, x=0.00, 0.03, 0.12.*

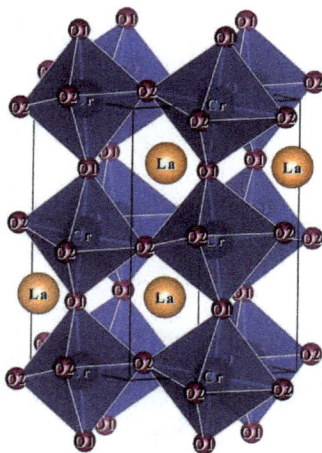

Figure 2.8 *The orthorhombic unit cell of $(La_{0.8}Ca_{0.2})(Cr_{0.87}Co_{0.1}Cu_{0.03})O_3$.*

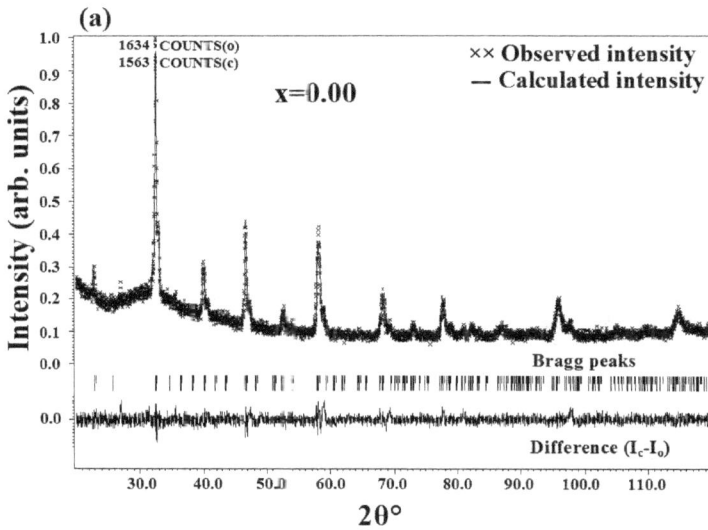

Figure 2.9 (a) *Fitted powder XRD profile for $(La_{0.8}Ca_{0.2})(Cr_{0.9-x}Co_{0.1}Cu_x)O_3$, x=0.00.*

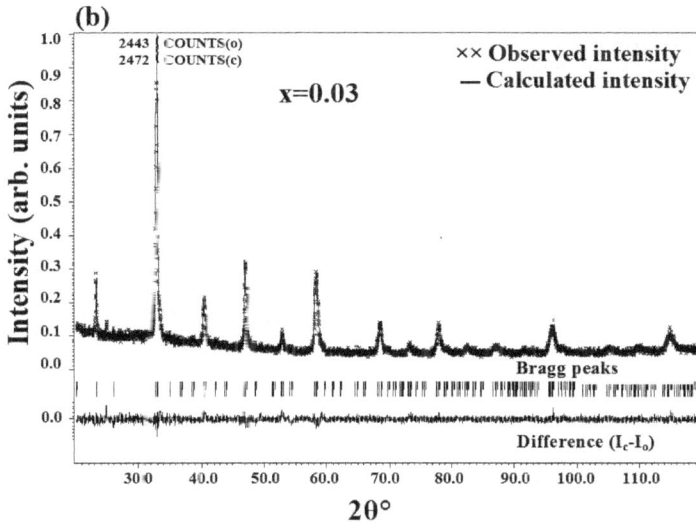

Figure 2.9 (b) *Fitted powder XRD profile for $(La_{0.8}Ca_{0.2})(Cr_{0.9-x}Co_{0.1}Cu_x)O_3$, x=0.03.*

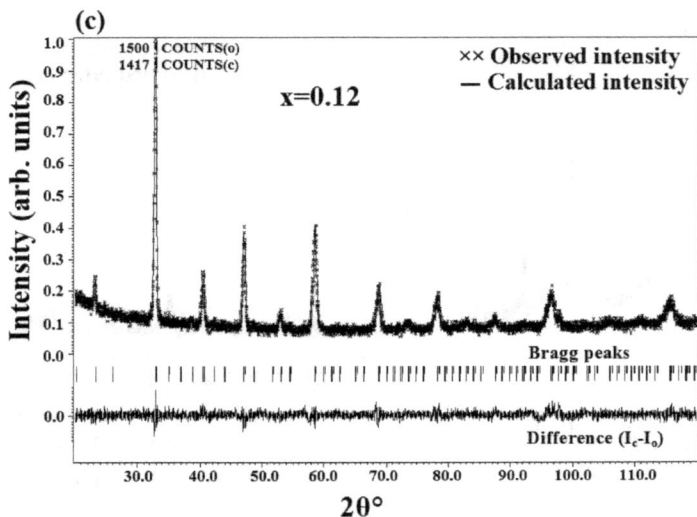

Figure 2.9 (c) *Fitted powder XRD profile for* $(La_{0.8}Ca_{0.2})(Cr_{0.9-x}Co_{0.1}Cu_x)O_3$, *x=0.12.*

Table 2.4 *Structural parameters for* $(La_{0.8}Ca_{0.2})(Cr_{0.9-x}Co_{0.1}Cu_x)O_3$, *x=0.00, 0.03, 0.12 through refinement of powder XRD data.*

Parameters	x=0.00	x=0.03	x=0.12
a (Å)	5.517(3)	5.484(9)	5.459(9)
b (Å)	7.792(8)	7.769(7)	7.724(5)
c (Å)	5.551(2)	5.511(3)	5.466(2)
$\alpha=\beta=\gamma$ (°)	90	90	90
Unit cell volume ($Å^3$)	238.68(8)	234.82(5)	230.23(4)
Density (gm/cc)	6.11(1)	6.18(6)	6.37(6)
R_p (%)	6.08	6.38	7.04
R_{obs} (%)	3.92	3.62	2.76
GOF	1.15	1.12	1.13
$F_{(000)}$	392	389	394

R_p- Reliability index for profile
R_{obs}- Reliability index for observed structure factors
GOF- Goodness of fit
$F_{(000)}$- Number of electrons in the unit cell

2.2.4 La$_{1-x}$Ca$_x$MnO$_3$ manganites

The observed powder X-ray diffraction patterns of La$_{1-x}$Ca$_x$MnO$_3$ (x=0.1, 0.2, 0.3, 0.4 and 0.5) manganite samples are shown in figure 2.10 (a). The enlarged XRD patterns corresponding to the (121) and (040) planes for all the Ca compositions shown in figure 2.10 (b). The distorted orthorhombic unit cell of La$_{1-x}$Ca$_x$MnO$_3$ obtained through VESTA [Momma and Izumi, 2008] is shown in figure 2.11.

The raw powder X-ray diffraction data of the synthesized manganite samples were refined for their structural parameters using the Rietveld technique [Rietveld, 1969]. The refinement was done with the software JANA 2006 [Petříček et al., 2014].

Figures 2.12 (a) - (e) show the fitted XRD profiles for La$_{1-x}$Ca$_x$MnO$_3$ (x=0.1, 0.2, 0.3, 0.4 and 0.5). The refined structural parameters and the reliability indices for all the Ca compositions are given in table 2.5.

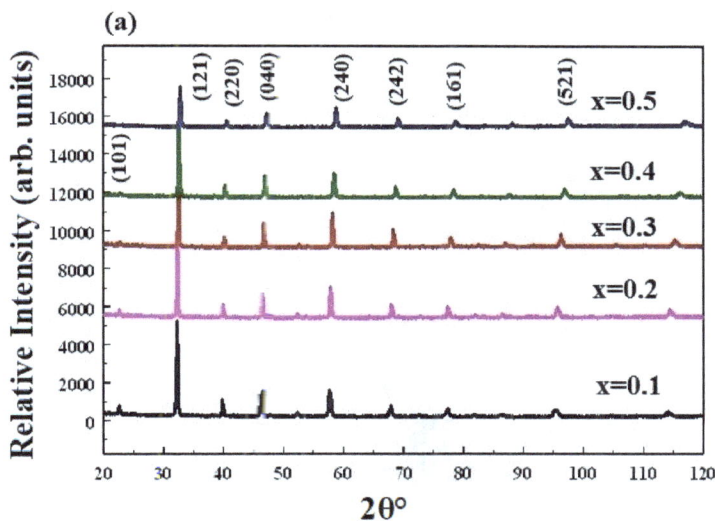

Figure 2.10 (a) *Observed X-ray powder diffractograms of La$_{1-x}$Ca$_x$MnO$_3$, x=0.1, 0.2, 0.3, 0.4, 0.5 ceramics.*

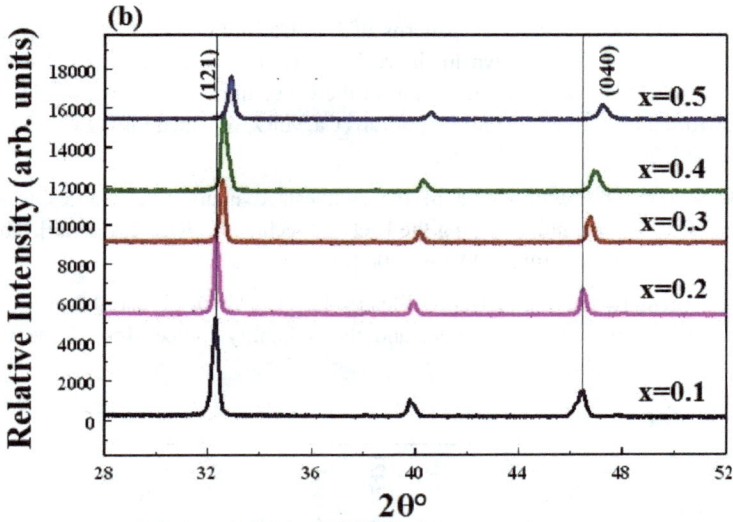

Figure 2.10 (b) *Enlarged X-ray diffraction pattern for (121) and (040) planes of La$_{1x}$Ca$_x$MnO$_3$, x=0.1, 0.2, 0.3, 0.4, 0.5 ceramics.*

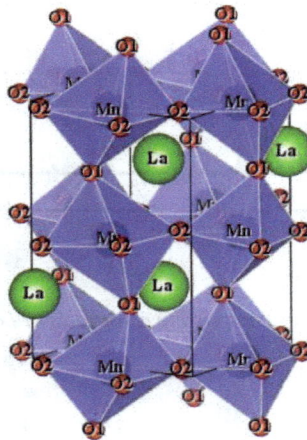

Figure 2.11 *The distorted orthorhombic unit cell of La$_{1-x}$Ca$_x$MnO$_3$.*

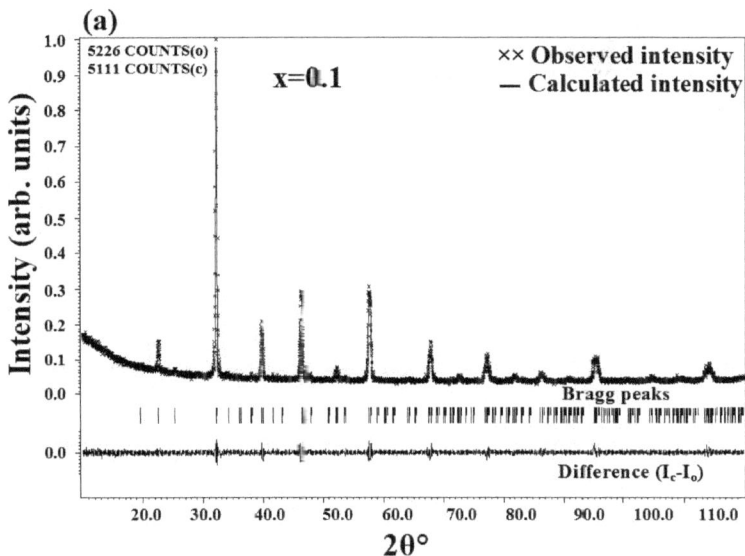

Figure 2.12 (a) Fitted powder XRD profile for La$_{1-x}$Ca$_x$MnO$_3$, x=0.1.

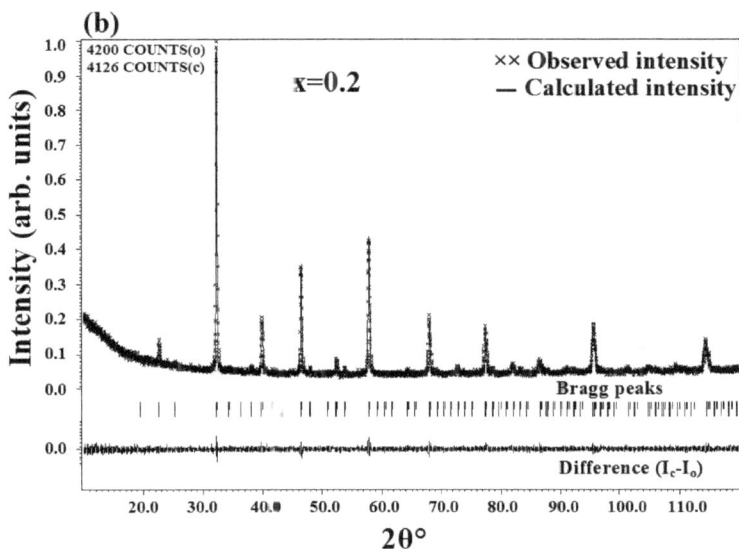

Figure 2.12 (b) Fitted powder XRD profile for La$_{1-x}$Ca$_x$MnO$_3$, x=0.2.

(c)

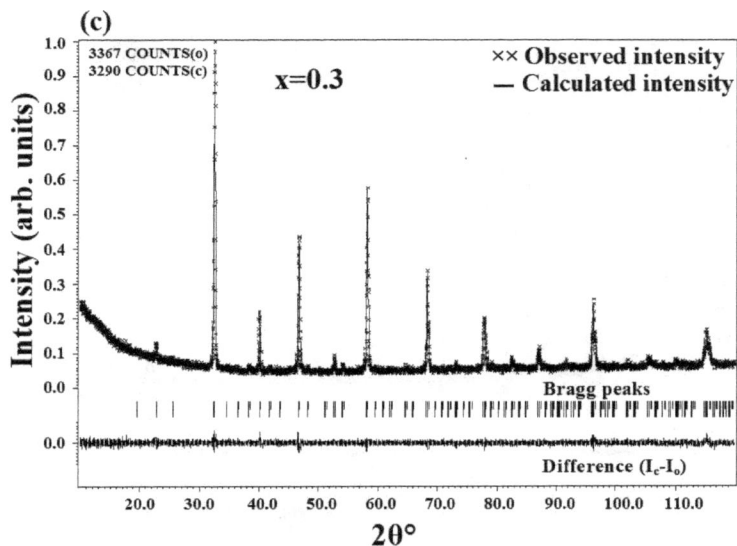

Figure 2.12 (c) Fitted powder XRD profile for $La_{1-x}Ca_xMnO_3$, x=0.3.

(d)

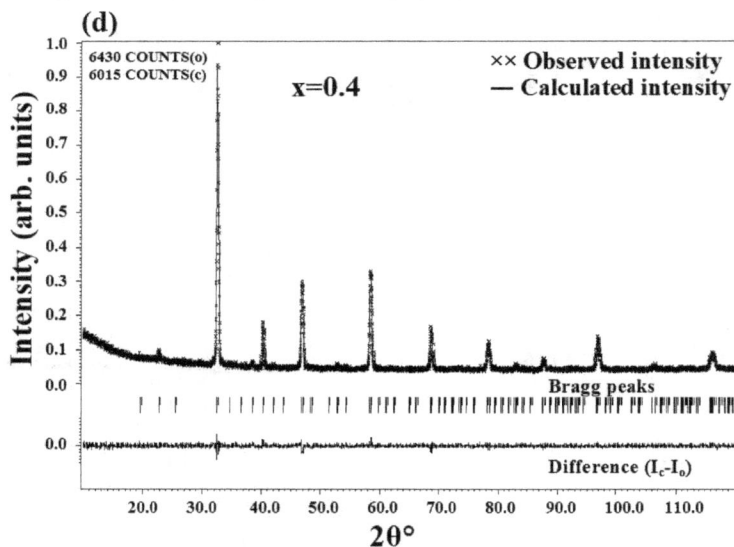

Figure 2.12 (d) Fitted powder XRD profile for $La_{1-x}Ca_xMnO_3$, x=0.4.

(e)

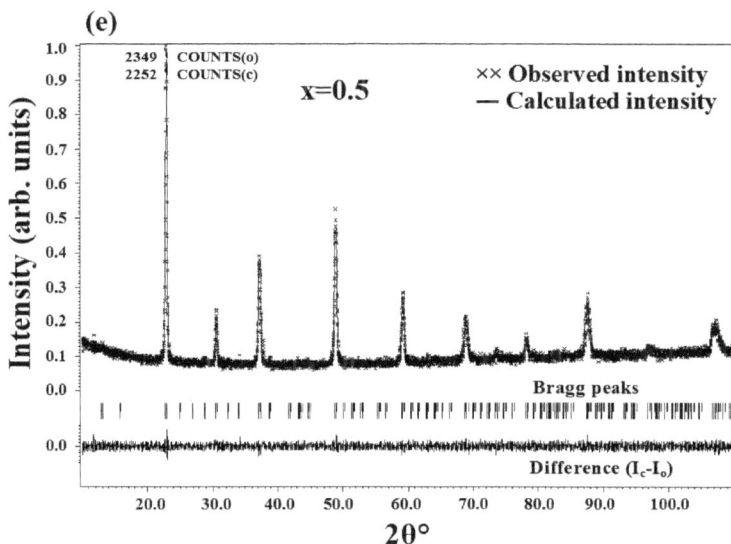

Figure 2.12 (e) Fitted powder XRD profile for La$_{1-x}$Ca$_x$MnO$_3$, x=0.5.

Table 2.5 *Structural parameters for La$_{1-x}$Ca$_x$MnO$_3$, x=0.1, 0.2, 0.3, 0.4, 0.5 through refinement of powder XRD data.*

Parameters	x=0.1	x=0.2	x=0.3	x=0.4	x=0.5
a (Å)	5.508(5)	5.482(9)	5.456(3)	5.431(12)	5.423(8)
b (Å)	7.747(5)	7.737(9)	7.710(3)	7.666(12)	7.632(8)
c (Å)	5.484(5)	5.483(9)	5.472(3)	5.444(12)	5.415(8)
α=β=γ (°)	90	90	90	90	90
Unit cell volume (Å^3)	234.07(6)	233.23(6)	230.24(4)	226.69(7)	224.18(5)
Density (gm/cc)	6.57(9)	6.32(2)	6.11(1)	5.92(3)	5.69(1)
R$_p$ (%)	5.07	4.90	4.96	4.28	5.16
R$_{obs}$ (%)	2.49	2.50	2.23	3.37	2.36
GOF	1.02	1.06	1.03	1.07	1.04
F$_{(000)}$	409	394	365	365	350

R$_p$- Reliability index for profile
R$_{obs}$- Reliability index for observed structure factors
GOF- Goodness of fit
F$_{(000)}$ - Number of electrons in the unit cell

2.2.5 $La_{1-x}Sr_xMnO_3$ manganites

The synthesized Sr doped lanthanum manganites have been characterized by powder XRD for their structural properties. The observed raw X-ray diffraction patterns of $La_{1-x}Sr_xMnO_3$, (x=0.3, 0.4 and 0.5) manganites are shown in figure 2.13 (a). The enlarged XRD patterns corresponding to the (110) and (202) planes for all the Sr compositions shown in figure 2.13 (b).

Powder XRD profile refinement of $La_{1-x}Sr_xMnO_3$ (x=0.3, 0.4 and 0.5) manganite samples have been carried out by Rietveld refinement [Rietveld, 1969] method using the software JANA 2006 [Petříček et al., 2014]. Figures 2.14 (a) - (c) show the fitted XRD profiles for $La_{1-x}Sr_xMnO_3$ (x=0.3, 0.4 and 0.5) manganites and the refined structural parameters and reliability indices are given in table 2.6.

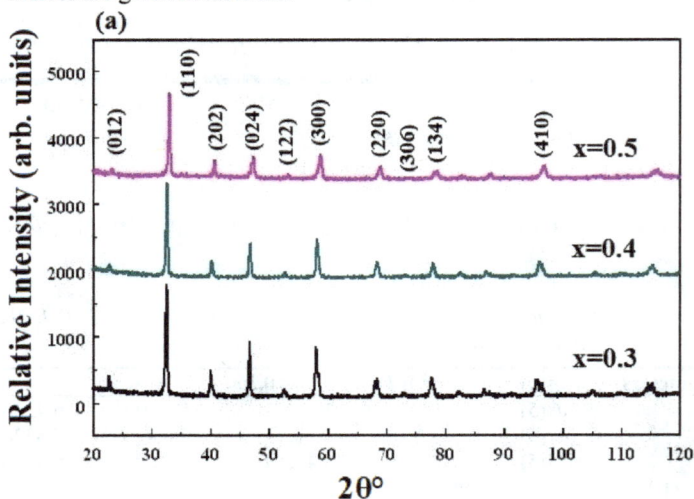

Figure 2.13 (a) *Observed powder X-ray diffractograms of $La_{1-x}Sr_xMnO_3$, x=0.3, 0.4 and 0.5.*

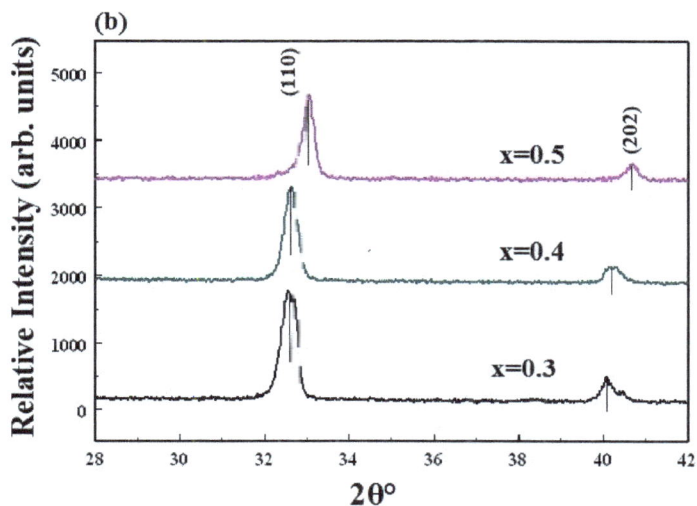

Figure 2.13 (b) Enlarged XRD peaks of (110) and (202) for La₁₋ₓSrₓMnO₃, x=0.3, 0.4 and 0.5.

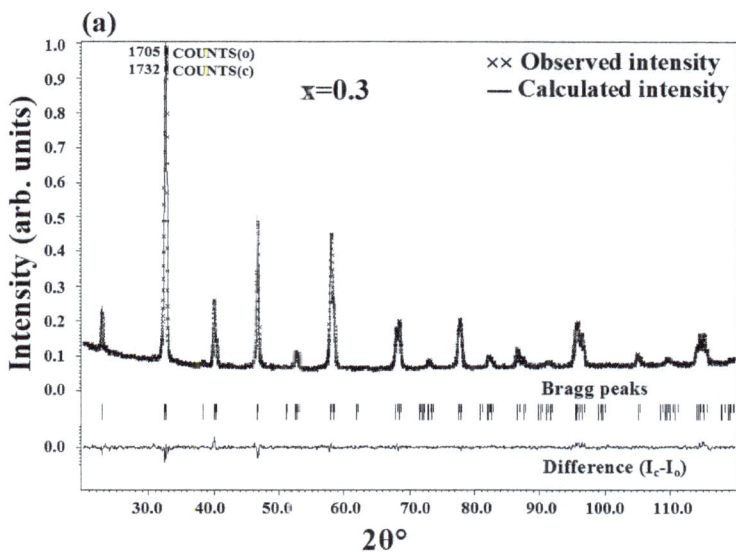

Figure 2.14 (a) Fitted powder XRD profile for La₁₋ₓSrₓMnO₃, x=0.3.

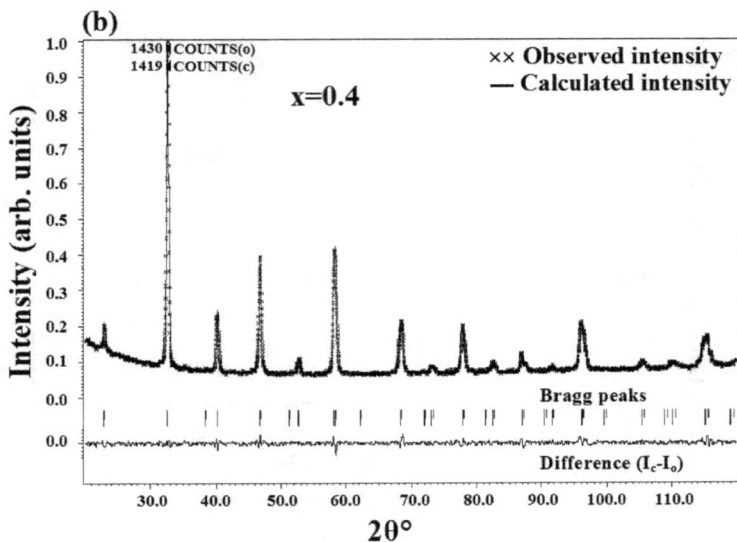

Figure 2.14 (b) Fitted powder XRD profile for $La_{1-x}Sr_xMnO_3$, $x=0.4$.

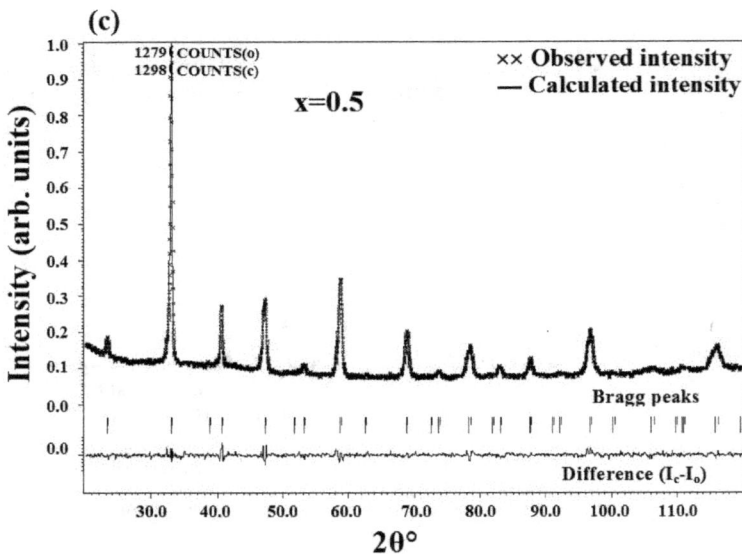

Figure 2.14 (c) Fitted powder XRD profile for $La_{1-x}Sr_xMnO_3$, $x=0.5$.

Table 2.6 *Structural parameters for $La_{1-x}Sr_xMnO_3$, x=0.3, 0.4 and 0.5 through refinement of powder XRD data.*

Parameters	x=0.3	x=0.4	x=0.5
Space group	$R\overline{3}c$	$R\overline{3}c$	$R\overline{3}c$
a (Å)	5.504(5)	5.495(11)	5.457(6)
b (Å)	5.504(5)	5.495(11)	5.457(6)
c (Å)	13.355(4)	13.378(3)	13.365(5)
$\alpha=\beta\neq\gamma$ (°)	90, 120	90, 120	90, 120
Unit cell volume (Å³)	350.37(3)	349.81(15)	344.83(4)
Density (gm/cc)	6.43(2)	6.30(2)	6.24(4)
R_p (%)	2.53	2.57	2.92
R_{obs} (%)	1.58	1.96	2.97
GOF	0.41	0.38	0.44
$F_{(000)}$	602	590	579

Rp- Reliability index for profile
Robs- Reliability index for observed structure factors
GOF- Goodness of fit
F(000)- Number of electrons in the unit cell

2.3 Morphological characterization and elemental confirmation - SEM/EDS

The SEM images of the co-doped lanthanum chromite samples were recorded at SAIF, Cochin, using the JEOL Model JSM - 6390LV scanning electron microscope. The SEM images for doped lanthanum manganite samples were recorded at Kalasalingam University, Krishnankoil, Tamil Nadu, India, using a Carl Zeiss EVO 18 scanning electron microscope. The EDS spectra for co-doped lanthanum chromite samples were obtained at SAIF, Cochin, using a JEOL Model JED – 2300 energy dispersive X-ray spectroscope. The EDS spectra for doped lanthanum manganite samples were obtained at Kalasalingam University, Krishnankoil, Tamil Nadu, India, using a Quantax 200 with X-flash-Bruker model energy dispersive X-ray spectroscope.

In this section, the SEM micrographs and the EDS spectra for all the doped lanthanum chromite and lanthanum manganite samples are presented. The atomic percentage and weight percentage of the various elements present in the samples for all the synthesized materials are tabulated and given in tables 2.7, 2.8, 2.9 and 2.10.

2.3.1 (Co, Mn) doped (La, Ca) based chromites - $(La_{0.8}Ca_{0.2})(Cr_{0.9-x}Co_{0.1}Mn_x)O_3$

The SEM images of the synthesized $(La_{0.8}Ca_{0.2})(Cr_{0.9-x}Co_{0.1}Mn_x)O_3$, (x=0.03, 0.06, 0.09 and 0.12) chromite samples were recorded for different magnifications (×1500, ×5000, ×10,000). Figures 2.15 (a) - (d) illustrate the SEM micrographs for the (Co, Mn) doped (La, Ca) based chromites for the magnification of ×10,000.

(a)

(b)

(c)

(d)

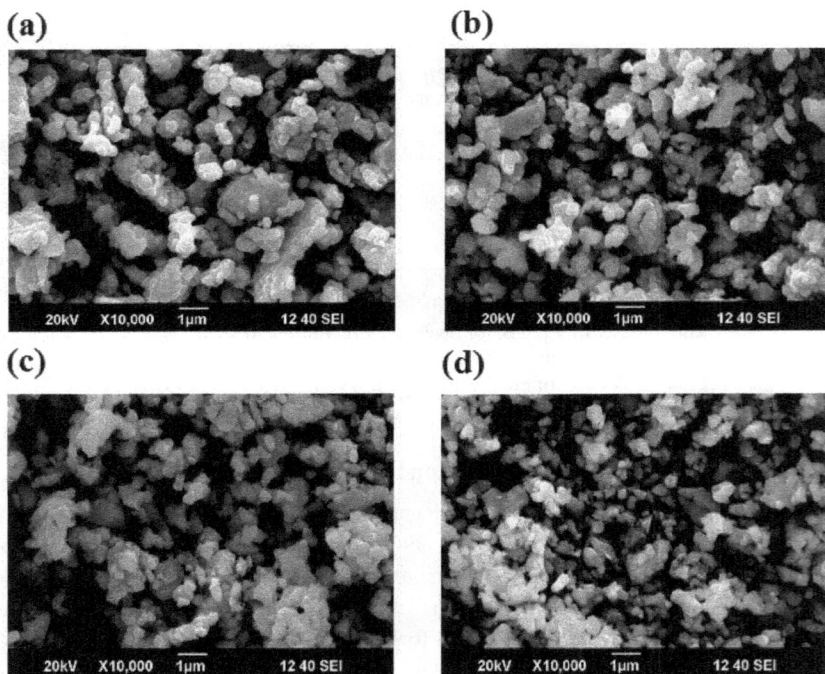

Figure 2.15 SEM images of $(La_{0.8}Ca_{0.2})(Cr_{0.9-x}Co_{0.1}Mn_x)O_3$, *(a)* x=0.03, *(b)* x=0.06, *(c)* x=0.09 and *(d)* x=0.12.

2.3.2 (Co, Fe) doped (La, Ca) based chromites - $(La_{0.8}Ca_{0.2})(Cr_{0.9-x}Co_{0.1}Fe_x)O_3$

Figures 2.16 (a) - (d) show the scanning electron microscope (SEM) images for the synthesized $(La_{0.8}Ca_{0.2})(Cr_{0.9-x}Co_{0.1}Fe_x)O_3$, x=0.03, 0.06, 0.09 and 0.12) samples for the magnification of ×10,000.

The EDS spectra for the synthesized $(La_{0.8}Ca_{0.2})(Cr_{0.9-x}Co_{0.1}Fe_x)O_3$, x=0.03, 0.06, 0.09 and 0.12) samples are presented in figures 2.17 (a) - (d). Table 2.7 gives the atomic percentage and weight percentage of the various elements present in the co-doped lanthanum chromite materials.

(a)

(b)

(c)

(d)

Figure 2.16 *SEM images of $(La_{0.8}Ca_{0.2})(Cr_{0.9-x}Co_{0.1}Fe_x)O_3$,* ***(a)*** *x=0.03,* ***(b)*** *x=0.06,* ***(c)*** *x=0.09 and* ***(d)*** *x=0.12.*

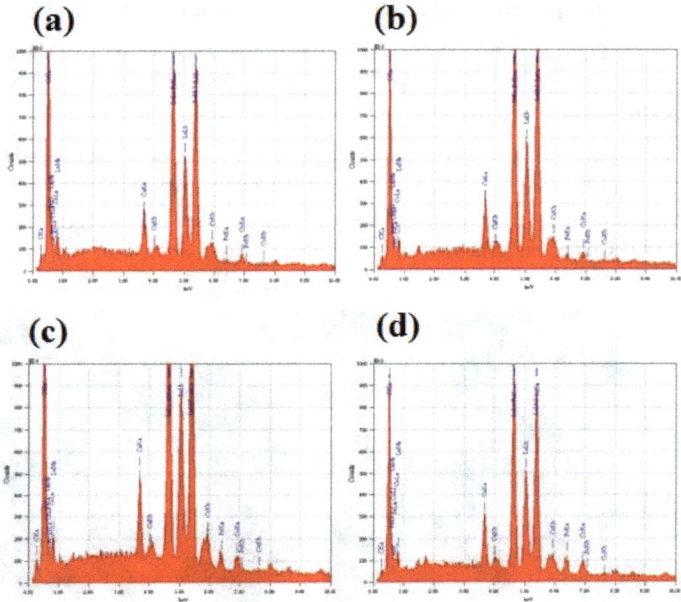

Figure 2.17 The EDS spectra of $(La_{0.8}Ca_{0.2})(Cr_{0.9-x}Co_{0.1}Fe_x)O_3$, **(a)** x=0.03, **(b)** x=0.06, **(c)** x=0.09 and **(d)** x=0.12.

Table 2.7 EDS elemental composition for $(La_{0.8}Ca_{0.2})(Cr_{0.9-x}Co_{0.1}Fe_x)O_3$, x=0.03, 0.06, 0.09, 0.12.

Elements	Atomic (%)				Weight (%)			
	x=0.03	x=0.06	x=0.09	x=0.12	x=0.03	x=0.06	x=0.09	x=0.12
La	27.55	29.89	29.04	28.91	63.50	64.49	62.96	62.66
Ca	4.40	5.16	5.67	6.17	2.93	3.21	3.55	3.86
Cr	23.37	26.25	26.22	23.36	20.16	21.20	21.28	18.95
Co	2.23	1.77	2.13	3.07	2.18	1.62	1.96	2.82
Fe	0.53	0.92	2.24	3.78	0.49	0.80	1.96	3.30
O	36.20	31.51	28.92	30.66	9.61	7.83	7.22	7.65

2.3.3 (Co, Cu) doped (La, Ca) based chromites - $(La_{0.8}Ca_{0.2})(Cr_{0.9-x}Co_{0.1}Cu_x)O_3$

The SEM micrographs of the prepared $(La_{0.8}Ca_{0.2})(Cr_{0.9-x}Co_{0.1}Cu_x)O_3$, (x=0.00, 0.03 and 0.12) samples are shown in figures 2.18 (a) - (c).

Figures 2.19 (a) - (c) show the EDS spectra for the synthesized co-doped chromite system $(La_{0.8}Ca_{0.2})(Cr_{0.9-x}Co_{0.1}Cu_x)O_3$, (x=0.00, 0.03 and 0.12). The elemental compositions in atomic and weight percentages are given in table 2.8.

(a)

(b)

(c)

Figure 2.18 SEM images of $(La_{0.8}Ca_{0.2})(Cr_{0.9-x}Co_{0.1}Cu_x)O_3$, *(a)* x=0.00, *(b)* x=0.03 and *(c)* x=0.12.

(a)

(b)

(c)

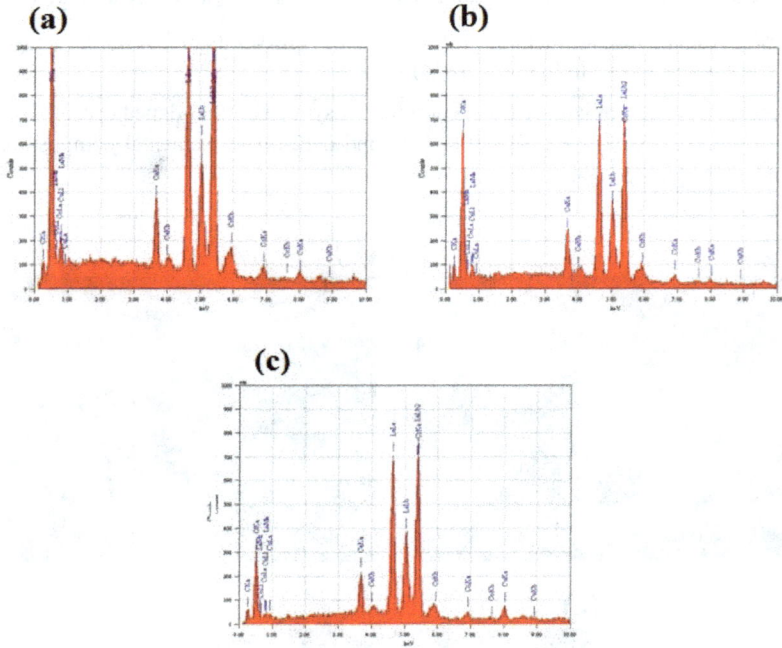

Figure 2.19 EDS spectra of $(La_{0.8}Ca_{0.2})(Cr_{0.9-x}Co_{0.1}Cu_x)O_3$, *(a)* x=0.00, *(b)* x=0.03 and *(c)* x=0.12.

Table 2.8 EDS elemental composition for $(La_{0.8}Ca_{0.2})(Cr_{0.9-x}Co_{0.1}Cu_x)O_3$, x=0.00, 0.03, 0.12.

Elements	Atomic (%)			Weight (%)		
	x=0.00	x=0.03	x=0.12	x=0.00	x=0.03	x=0.12
La	24.93	26.93	31.47	59.85	62.26	62.86
Ca	5.79	6.27	5.89	4.01	4.18	3.4
Cr	23.5	23.87	29.17	20.66	20.66	21.81
Co	2.25	1.88	1.9	2.29	1.84	1.61
Cu	-	1.24	5.39	-	1.31	4.93
O	33.97	26.87	15.30	9.39	7.16	3.52

2.3.4 La$_{1-x}$Ca$_x$MnO$_3$ manganites

The SEM micrographs of the synthesized calcium doped lanthanum manganite La$_{1-x}$Ca$_x$MnO$_3$, (x=0.1, 0.2, 0.3, 0.4 and 0.5) samples with magnification of ×10,000 are presented in figures 2.20 (a) - (e).

The EDS spectra for the calcium doped lanthanum manganite La$_{1-x}$Ca$_x$MnO$_3$, (x=0.1, 0.2, 0.3, 0.4 and 0.5) samples are shown in figures 2.21 (a) - (e). Table 2.9 gives the atomic and weight percentages of La, Ca, Mn and O atoms present in the sample.

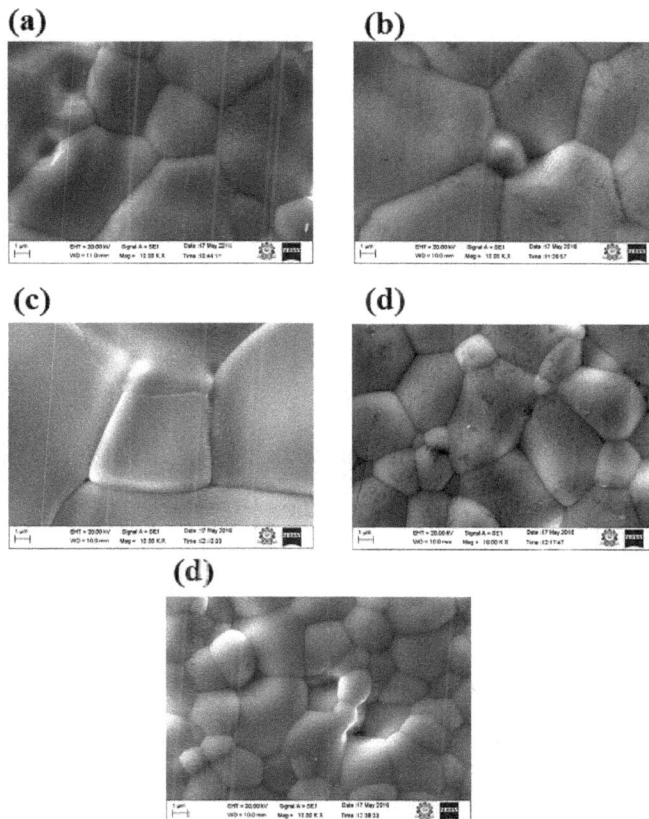

Figure 2.20 SEM images of La$_{1-x}$Ca$_x$MnO$_3$, **(a)** x=0.1, **(b)** x=0.2, **(c)** x=0.3, **(d)** x=0.4 and **(e)** x=0.5 ceramics.

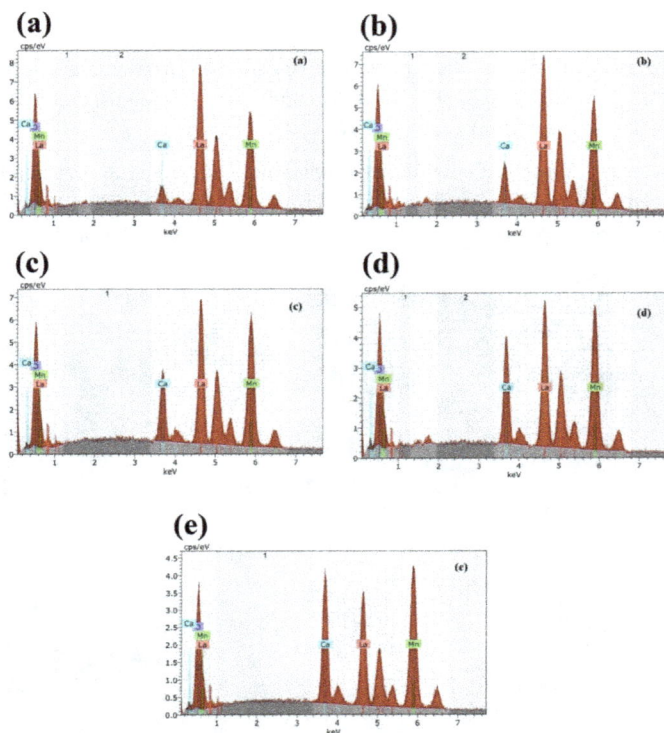

Figure 2.21 EDS spectra of $La_{1-x}Ca_xMnO_3$, **(a)** x=0.1, **(b)** x=0.2, **(c)** x=0.3, **(d)** x=0.4 and **(e)** x=0.5 ceramics.

Table 2.9 EDS elemental composition for $La_{1-x}Ca_xMnO_3$, x=0.1, 0.2, 0.3, 0.4, 0.5.

Samples	Atomic (%)				Weight (%)			
	La	Ca	Mn	O	La	Ca	Mn	O
x=0.1	16.94	2.18	19.16	61.71	52.51	1.95	23.50	22.04
x=0.2	15.02	4.22	18.16	62.59	49.04	3.98	23.45	23.53
x=0.3	13.15	6.44	19.09	61.22	44.40	6.27	25.48	23.84
x=0.4	11.09	8.60	18.02	62.30	39.79	8.90	25.57	25.75
x=0.5	9.22	10.95	17.87	61.96	34.67	11.89	26.59	26.85

2.3.5 La$_{1-x}$Sr$_x$MnO$_3$ manganites

The SEM micrographs of strontium doped lanthanum manganite La$_{1-x}$Sr$_x$MnO$_3$, (x=0.3, 0.4 and 0.5) samples taken with ×15,000 magnification are shown in figures 2.22 (a) - (c).

Figures 2.23 (a) - (c) show the EDS spectra for the strontium doped lanthanum manganite La$_{1-x}$Sr$_x$MnO$_3$, (x=0.3, 0.4 and 0.5) samples and the atomic and weight percentages of La, Sr, Mn and O atoms present in the synthesized samples are given in table 2.10.

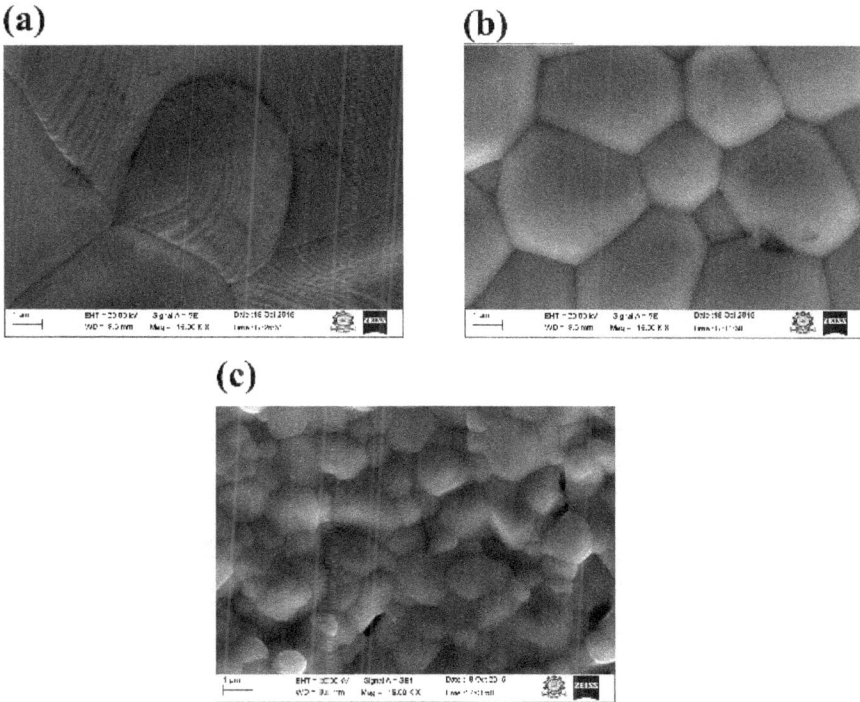

(a)

(b)

(c)

Figure 2.22 SEM images of La$_{1-x}$Sr$_x$MnO$_3$, *(a)* x=0.3, *(b)* x=0.4 and *(c)* x=0.5.

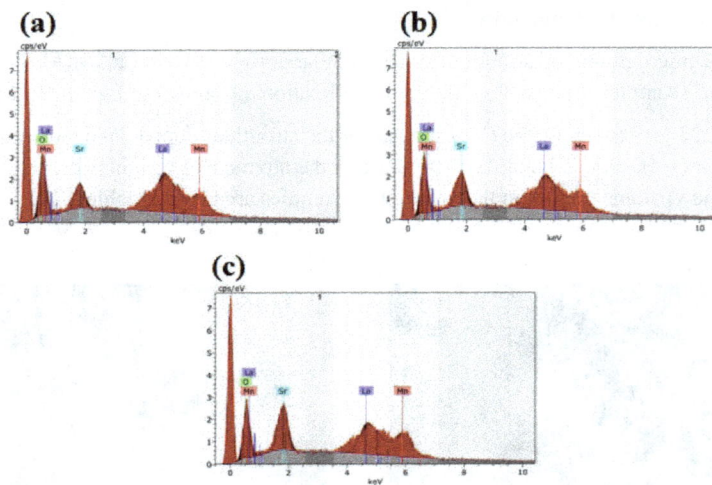

Figure 2.23 *EDS spectra of La$_{1-x}$Sr$_x$MnO$_3$,* **(a)** *x=0.3,* **(b)** *x=0.4 and* **(c)** *x=0.5.*

Table 2.10 *EDS elemental composition for La$_{1-x}$Sr$_x$MnO$_3$, x=0.3, 0.4 and 0.5.*

Samples	Atomic (%)				Weight (%)			
	La	Sr	Mn	O	La	Sr	Mn	O
x=0.3	13.41	5.72	12.85	68.03	44.79	12.05	16.98	26.18
x=0.4	11.92	7.51	12.84	67.73	40.37	16.03	17.19	26.41
x=0.5	10.97	10.49	15.14	63.41	35.53	21.43	19.39	23.66

2.4   Optical characterization - UV-visible absorption spectra

The UV-visible absorption spectra for the doped lanthanum chromite and lanthanum manganite samples were recorded at SAIF, Cochin, using a Cary 5000 (Varian, Germany) UV-visible spectrophotometer. The UV-visible absorption spectra were taken in the range from 200 nm to 2000 nm. The optical band gap was estimated using Wood and Tauc's relation $\alpha h v = A (h v - E_g)^n$ [Wood and Tauc, 1972].

In this section, the UV-visible absorption spectra and Tauc plot [Wood and Tauc, 1972] for all the synthesized samples are presented. The optical band gap values for all the prepared samples are tabulated and given in tables 2.11, 2.12, 2.13, 2.14 and 2.15.

2.4.1   (Co, Mn) doped (La, Ca) based chromites - $(La_{0.8}Ca_{0.2})(Cr_{0.9-x}Co_{0.1}Mn_x)O_3$

The UV-visible absorption spectra for the synthesized $(La_{0.8}Ca_{0.2})(Cr_{0.9-x}Co_{0.1}Mn_x)O_3$, (x=0.03, 0.06, 0.09 and 0.12) chromite samples are shown in figure 2.24. The Tauc plot, (hv) vs $(\alpha h v)^2$ [Wood and Tauc, 1972] to estimate the optical band gap of the prepared samples is shown in figure 2.25.

The optical band gap values for the (Co, Mn) doped (La, Ca) based chromites are given in table 2.11.

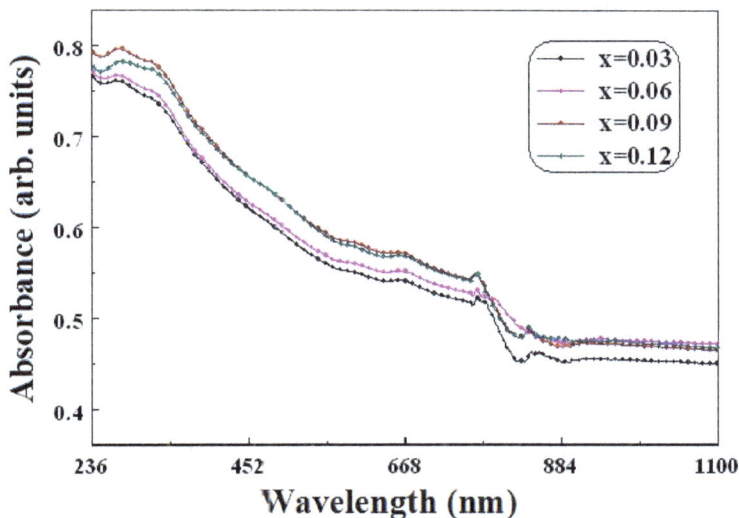

Figure 2.24 *UV-visible absorption spectra of $(La_{0.8}Ca_{0.2})(Cr_{0.9-x}Co_{0.1}Mn_x)O_3$, x=0.03, 0.06, 0.09, 0.12.*

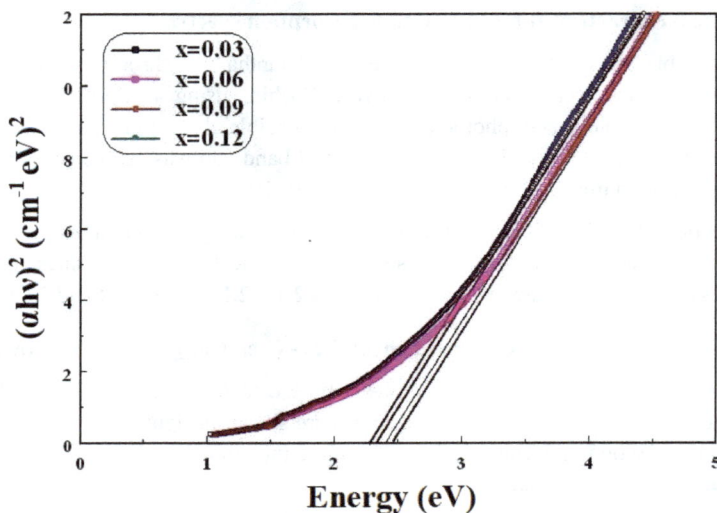

Figure 2.25 Tauc plot for $(La_{0.8}Ca_{0.2})(Cr_{0.9-x}Co_{0.1}Mn_x)O_3$, x=0.03, 0.06, 0.09, 0.12.

Table 2.11 Optical band gap values for $(La_{0.8}Ca_{0.2})(Cr_{0.9-x}Co_{0.1}Mn_x)O_3$, x=0.03, 0.06, 0.09, 0.12 from UV-vis analysis.

Samples	Energy gap (eV)
x=0.03	2.464
x=0.06	2.399
x=0.09	2.270
x=0.12	2.335

2.4.2 (Co, Fe) doped (La, Ca) based chromites - $(La_{0.8}Ca_{0.2})(Cr_{0.9-x}Co_{0.1}Fe_x)O_3$

Figure 2.26 shows the UV-visible absorption spectra for the synthesized $(La_{0.8}Ca_{0.2})(Cr_{0.9-x}Co_{0.1}Fe_x)O_3$, (x=0.03, 0.06, 0.09 and 0.12) samples. The optical band gap has been estimated from the Tauc plot, (hv) vs $(αhv)^2$ [Wood and Tauc, 1972] which is shown in figure 2.27. The direct band gap values are presented in table 2.12 for various concentration of Fe.

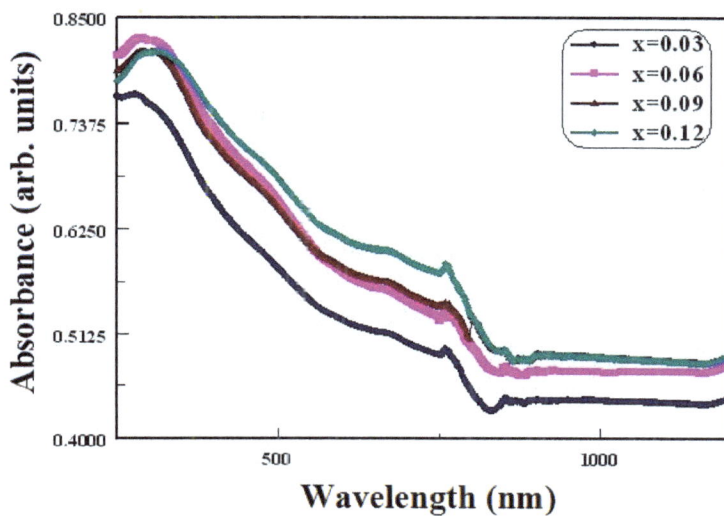

Figure 2.26 UV-visible absorption spectra for $(La_{0.8}Ca_{0.2})(Cr_{0.9-x}Co_{0.1}Fe_x)O_3$, x=0.03, 0.06, 0.09, 0.12.

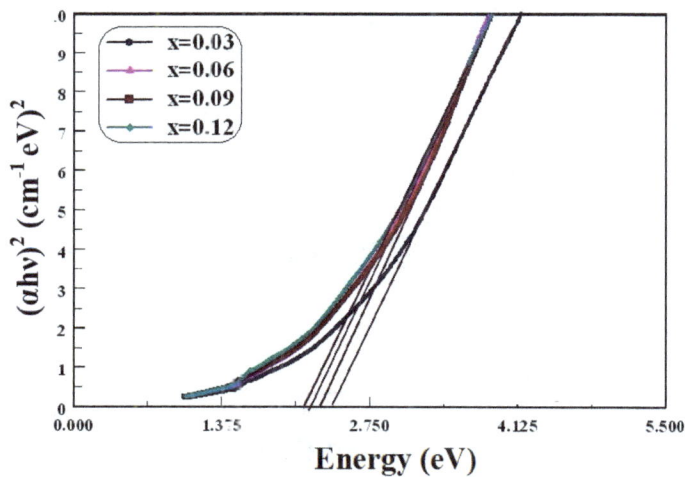

Figure 2.27 Tauc plot for $(La_{0.8}Ca_{0.2})(Cr_{0.9-x}Co_{0.1}Fe_x)O_3$, x=0.03, 0.06, 0.09, 0.12.

Table 2.12 Optical band gap values for $(La_{0.8}Ca_{0.2})(Cr_{0.9-x}Co_{0.1}Fe_x)O_3$, $x=0.03$, 0.06, 0.09, 0.12 from UV-vis analysis

Samples	Energy gap (eV)
x=0.03	2.405
x=0.06	2.292
x=0.09	2.186
x=0.12	2.135

2.4.3 (Co, Cu) doped (La, Ca) based chromites - $(La_{0.8}Ca_{0.2})(Cr_{0.9-x}Co_{0.1}Cu_x)O_3$

The UV-visible absorption spectra for the prepared $(La_{0.8}Ca_{0.2})(Cr_{0.9-x}Co_{0.1}Cu_x)O_3$, (x=0.00, 0.03, 0.09 and 0.12) chromite sample are shown in figure 2.28. Tauc plot [Wood and Tauc, 1972] is drawn for synthesized materials, by taking photon energy (hν) along the x-axis and $(αhν)^2$ along the y-axis and is shown in figure 2.29. The optical band gap values obtained from figure 2.29 are tabulated in table 2.13.

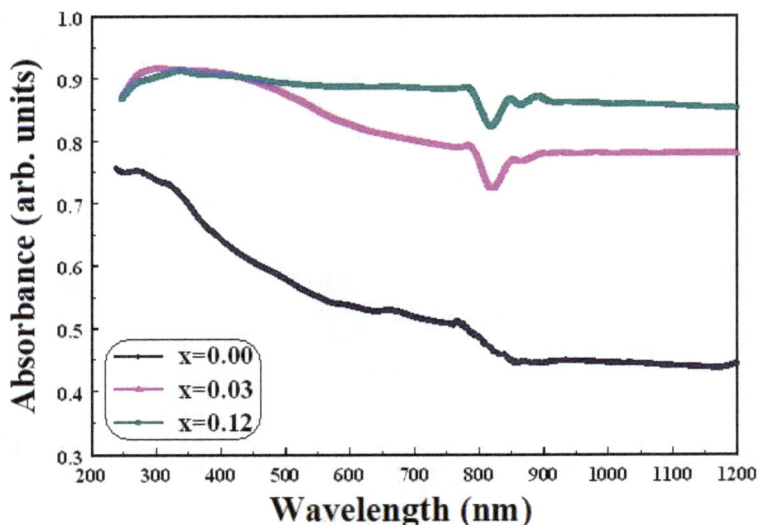

Figure 2.28 UV-visible absorption spectra of $(La_{0.8}Ca_{0.2})(Cr_{0.9-x}Co_{0.1}Cu_x)O_3$, $x=0.00$, 0.03 and 0.12.

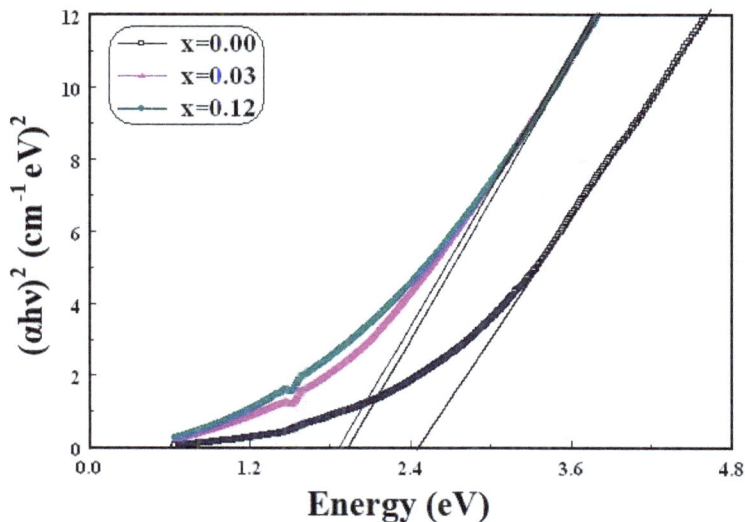

Figure 2.29 Tauc plot for $(La_{0.3}Ca_{0.2})(Cr_{0.9-x}Co_{0.1}Cu_x)O_3$, x=0.00, 0.03 and 0.12.

Table 2.13 *Optical band gap values for $(La_{0.8}Ca_{0.2})(Cr_{0.9-x}Co_{0.1}Cu_x)O_3$, x=0.00, 0.03, 0.12 from UV-vis analysis.*

Samples	Energy gap (eV)
x=0.00	2.448
x=0.03	1.934
x=0.12	1.859

2.4.4 La$_{1-x}$Ca$_x$MnO$_3$ manganites

Figure 2.30 shows the UV-visible absorption spectra for the synthesized calcium doped lanthanum manganite samples La$_{1-x}$Ca$_x$MnO$_3$ (x=0.1, 0.2, 0.3, 0.4 and 0.5). Using the Tauc relation [Wood and Tauc, 1972], a graph between (hv) and (αhv)^2 is drawn as shown in figure 2.31 which estimates the optical band gap of the synthesized samples. The optical band gap values for La$_{1-x}$Ca$_x$MnO$_3$ (x=0.1, 0.2, 0.3, 0.4 and 0.5) are given in table 2.14.

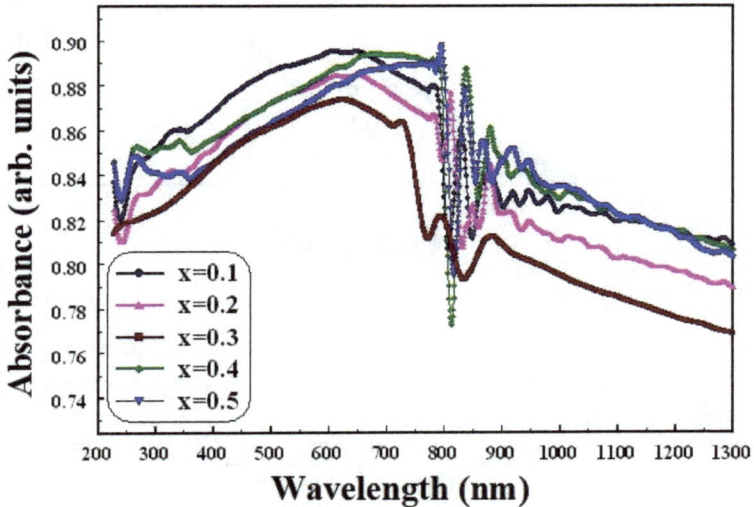

Figure 2.30 *UV-visible absorption spectra of La$_{1-x}$Ca$_x$MnO$_3$, x=0.1, 0.2, 0.3, 0.4, 0.5 ceramics.*

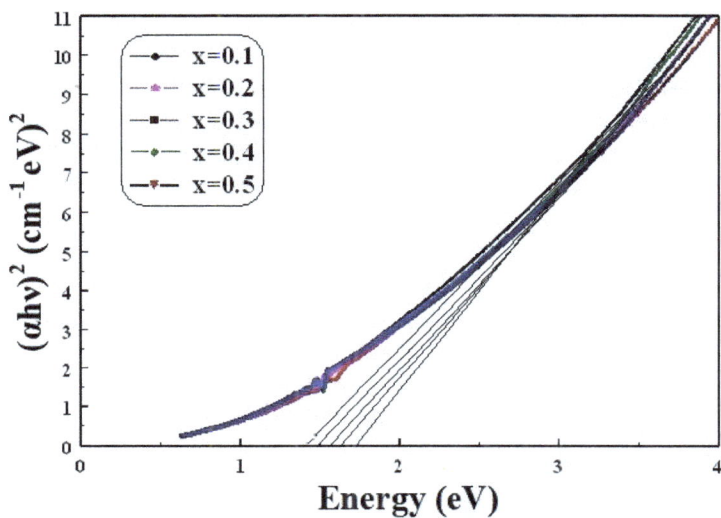

Figure 2.31 *Tauc plot for* $La_{1-x}Ca_xMnO_3$, *x=0.1, 0.2, 0.3, 0.4, 0.5 ceramics.*

Table 2.14 *Optical band gap values for* $La_{1-x}Ca_xMnO_3$ *x=0.1, 0.2, 0.3, 0.4, 0.5 from UV-vis analysis.*

Samples	Energy gap (eV)
x=0.1	1.730
x=0.2	1.626
x=0.3	1.571
x=0.4	1.476
x=0.5	1.411

2.4.5 La$_{1-x}$Sr$_x$MnO$_3$ manganites

The UV-visible absorption spectra of La$_{1-x}$Sr$_x$MnO$_3$, (x=0.3, 0.4 and 0.5) manganite samples are shown in figure 2.32. Figure 2.33 shows the Tauc plot [Wood and Tauc, 1972] ((hν) vs (αhν)$^{1/2}$), which estimates the optical band gaps of the prepared strontium doped lanthanum manganite samples. The indirect energy band gap values for La$_{1-x}$Sr$_x$MnO$_3$, (x=0.3, 0.4 and 0.5) manganite samples are tabulated in table 2.15.

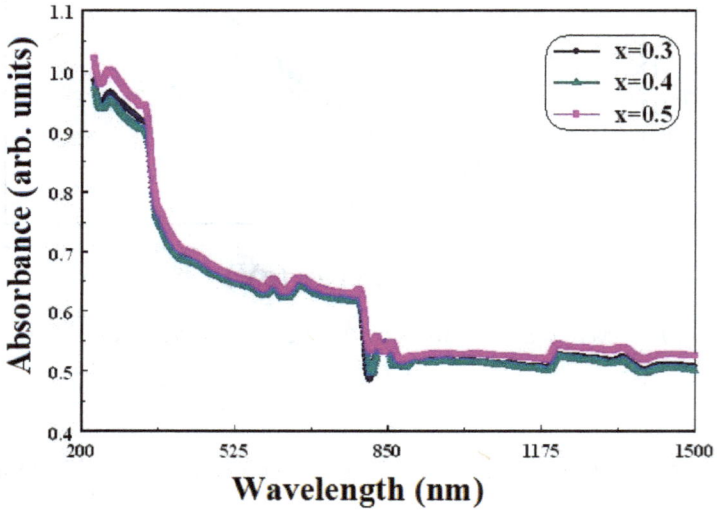

Figure 2.32 *UV-visible absorption spectra of La$_{1-x}$Sr$_x$MnO$_3$, x=0.3, 0.4 and 0.5.*

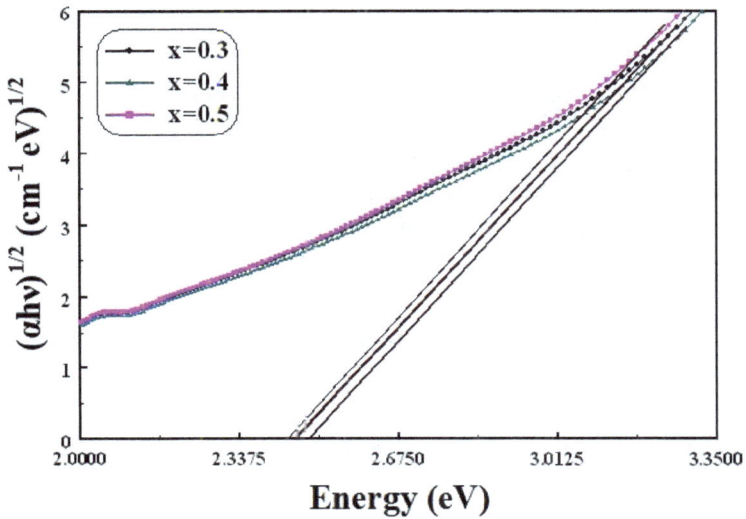

Figure 2.33 *Tauc plot for* $La_{1-x}Sr_xMnO_3$, *x=0.3, 0.4 and 0.5.*

Table 2.15 *Optical band gap values of* $La_{1-x}Sr_xMnO_3$, *x= 0.3, 0.4 and 0.5 from UV-vis analysis.*

Samples	Energy gap (eV)
x=0.3	2.457
x=0.4	2.487
x=0.5	2.442

2.5 Magnetic characterization – Vibrating sample magnetometry

The synthesized doped lanthanum chromite and lanthanum manganite samples were characterized for their magnetic properties using a Lakeshore VSM 7410 model vibrating sample magnetometer (VSM) at SAIF, IIT Madras, Chennai, India.

In this section, the room temperature M-H curves for all the samples are presented. The low temperature (20 K) M-H curve for Ca doped $LaMnO_3$ is also presented. The magnetic parameters such as saturation magnetization (M_s), remnant magnetization (M_r) and coercive field (H_c) values for all the samples are tabulated and given in tables 2.16, 2.17, 2.18, 2.19, 2.20 and 2.21.

2.5.1 (Co, Mn) doped (La, Ca) based chromites - $(La_{0.8}Ca_{0.2})(Cr_{0.9-x}Co_{0.1}Mn_x)O_3$

The room temperature M-H curves for the co-doped lanthanum chromite system $(La_{0.8}Ca_{0.2})(Cr_{0.9-x}Co_{0.1}Mn_x)O_3$, (x=0.03, 0.06, 0.09 and 0.12) are shown in figure 2.34. The magnetic parameters such as saturation magnetization (M_s), remnant magnetization (M_r) and coercive field (H_c) values are given in table 2.16.

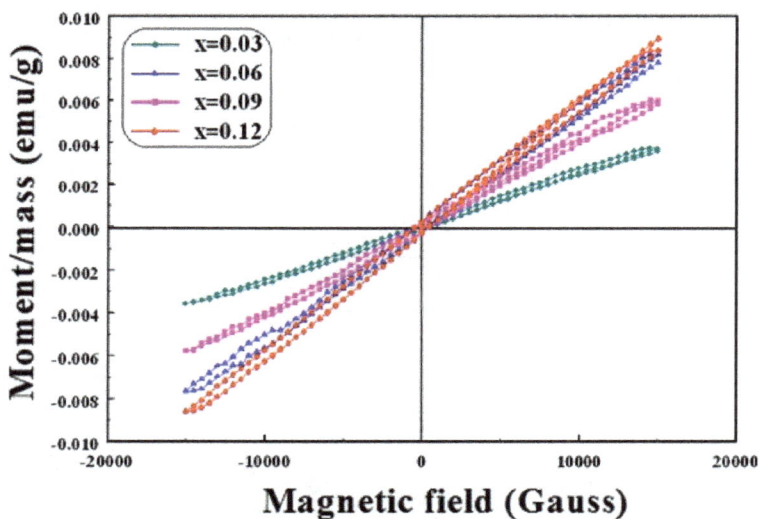

Figure 2.34 M-H curves for $(La_{0.8}Ca_{0.2})(Cr_{0.9-x}Co_{0.1}Mn_x)O_3$, x=0.03, 0.06, 0.09 and 0.12.

Table 2.16 *Magnetic parameters for $(La_{0.8}Ca_{0.2})(Cr_{0.9-x}Co_{0.1}Mn_x)O_3$, x=0.03, 0.06, 0.09, 0.12.*

Parameters	x=0.03	x=0.06	x=0.09	x=0.12
$M_s \times 10^{-3}$ (emu g^{-1})	3.64	7.90	5.89	8.76
H_c (G)	348.01	379.29	444.63	430.95
$M_r \times 10^{-3}$ (emu g^{-1})	0.106	0.226	0.205	0.254

M_s - Saturation magnetization
H_c - Coercive field
M_r - Remnant magnetization

2.5.2 (Co, Fe) doped (La, Ca) based chromites - $(La_{0.8}Ca_{0.2})(Cr_{0.9-x}Co_{0.1}Fe_x)O_3$

The M-H loops recorded for the $(La_{0.8}Ca_{0.2})(Cr_{0.9-x}Co_{0.1}Fe_x)O_3$, (x=0.03, 0.06, 0.09 and 0.12) samples at 300 K are shown in figure 2.35. The magnetic parameters are listed in table 2.17.

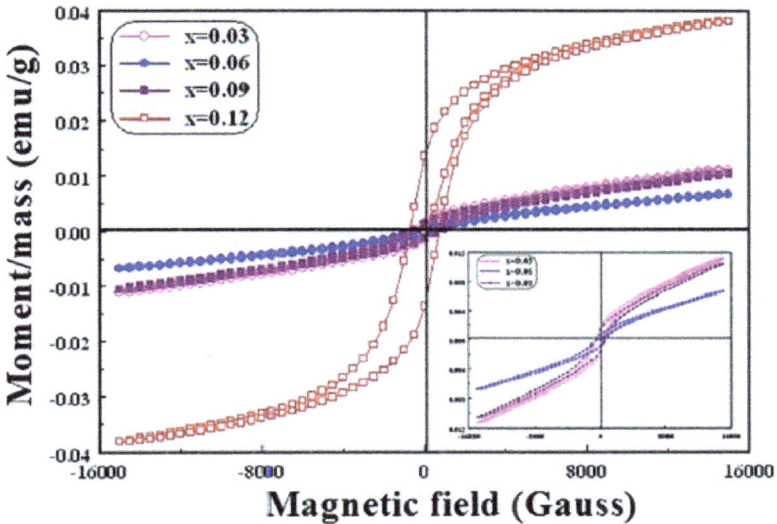

Figure 2.35 *M-H curves for $(La_{0.8}Ca_{0.2})(Cr_{0.9-x}Co_{0.1}Fe_x)O_3$, x=0.03, 0.06, 0.09 and 0.12.*

Table 2.17 *Magnetic parameters for $(La_{0.8}Ca_{0.2})(Cr_{0.9-x}Co_{0.1}Fe_x)O_3$, $x=0.03$, 0.06, 0.09, 0.12.*

Parameters	x=0.03	x=0.06	x=0.09	x=0.12
$M_s \times 10^{-3}$ (emu g^{-1})	11.20	6.71	10.47	38.12
H_c (G)	502.30	567.80	641.50	724.24
$M_r \times 10^{-3}$ (emu g^{-1})	1.589	0.695	1.416	13.649

M_s - Saturation magnetization
H_c - Coercive field
M_r - Remnant magnetization

2.5.3 (Co, Cu) doped (La, Ca) based chromites - $(La_{0.8}Ca_{0.2})(Cr_{0.9-x}Co_{0.1}Cu_x)O_3$

The room temperature M-H curves for the $(La_{0.8}Ca_{0.2})(Cr_{0.9-x}Co_{0.1}Cu_x)O_3$, ($x=0.00$, 0.03 and 0.12) samples are shown in figure 2.36. The magnetic parameters such as M_s, M_r and H_c values are given in table 2.18.

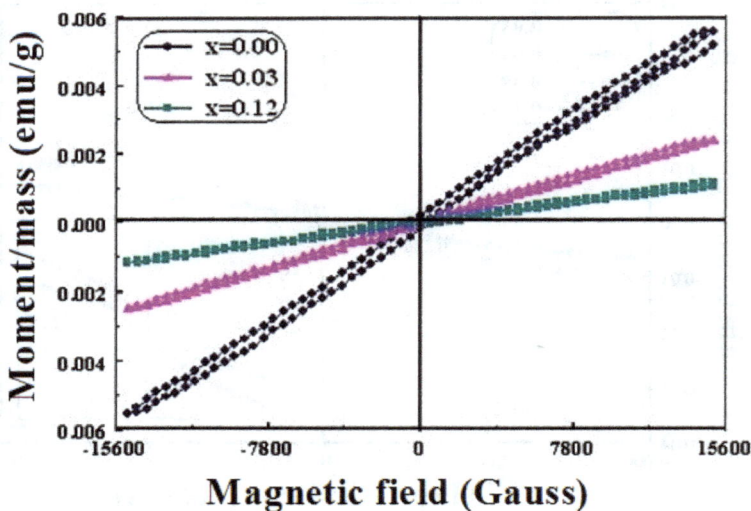

Figure 2.36 *M-H curves for $(La_{0.8}Ca_{0.2})(Cr_{0.9-x}Co_{0.1}Cu_x)O_3$, $x=0.00$, 0.03 and 0.12.*

Table 2.18 *Magnetic parameters for* $(La_{0.8}Ca_{0.2})(Cr_{0.9-x}Co_{0.1}Cu_x)O_3$, *x=0.00, 0.03, 0.12.*

Parameters	x=0.00	x=0.03	x=0.12
$M_s \times 10^{-3}$ (emu g^{-1})	5.55	2.43	1.14
H_c (G)	397.86	428.95	634.97
$M_r \times 10^{-3}$ (emu g^{-1})	179.43	70.081	36.81

M_s - Saturation magnetization
H_c - Coercive field
M_r - Remnant magnetization

2.5.4 La$_{1-x}$Ca$_x$MnO$_3$ manganites

The synthesized La$_{1-x}$Ca$_x$MnO$_3$ (x=0.1, 0.2, 0.3, 0.4 and 0.5) samples have been analyzed for their magnetic properties at 20 K and 300 K. The M-H loops recorded at 20 K are shown in figure 2.37 and the magnetic parameters for 20 K are given in table 2.19. Figure 2.38 shows the M-H loops recorded at 300 K for the synthesized Ca doped LaMnO$_3$ samples and the magnetic parameters are presented in table 2.20.

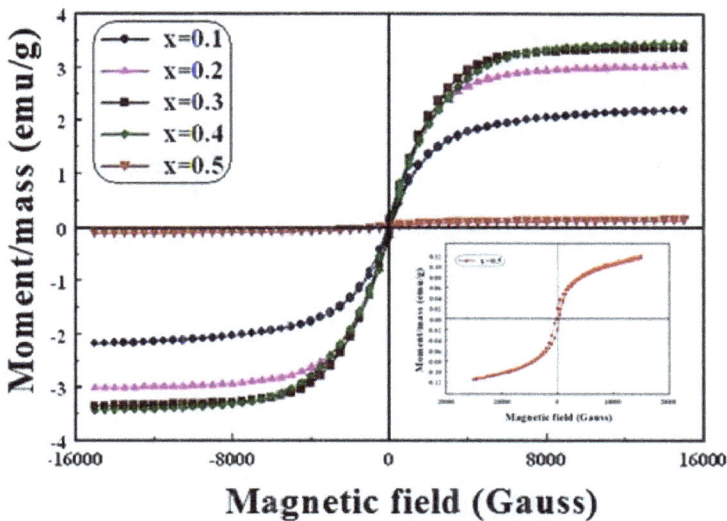

Figure 2.37 *M-H curves for* La$_{1-x}$Ca$_x$MnO$_3$, *x=0.1, 0.2, 0.3, 0.4, 0.5 at 20 K.*

Table 2.19 *Magnetic parameters for La$_{1-x}$Ca$_x$MnO$_3$, x=0.1, 0.2, 0.3, 0.4, 0.5 at 20 K.*

Parameters	x=0.1	x=0.2	x=0.3	x=0.4	x=0.5
M$_s$ (emu g^{-1})	2.189	3.015	3.354	3.438	0.117
H$_c$ (G)	99.42	56.36	66.88	127.93	356.30
M$_r$×10^{-3} (emu g^{-1})	107.41	86.57	99.79	196.14	19.94

M$_s$ - Saturation magnetization
H$_c$ - Coercive field
M$_r$ - Remnant magnetization

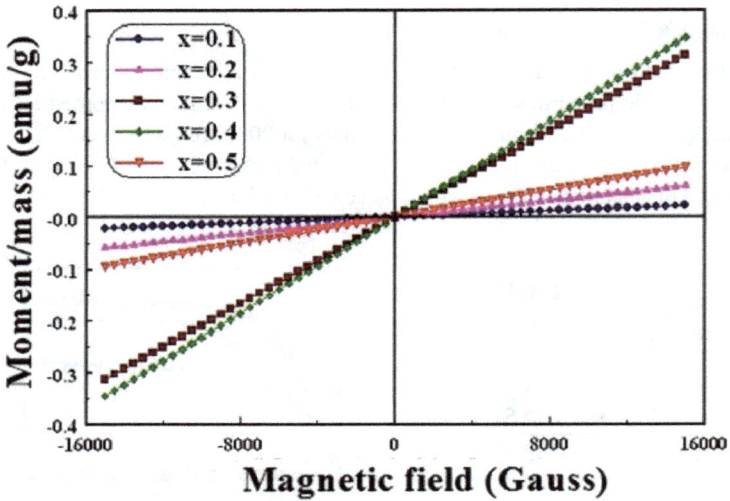

Figure 2.38 *M-H curves for La$_{1-x}$Ca$_x$MnO$_3$, x=0.1, 0.2, 0.3, 0.4, 0.5 at 300 K.*

Table 2.20 *Magnetic parameters for La$_{1-x}$Ca$_x$MnO$_3$, x=0.1, 0.2, 0.3, 0.4, 0.5 at 300 K.*

Parameters	x=0.1	x=0.2	x=0.3	x=0.4	x=0.5
M$_s$ ×10^{-3} (emu g^{-1})	21.69	58.35	313.88	347.30	94.79
H$_c$ (G)	37.21	38.77	32.12	31.54	40.47
M$_r$×10^{-6} (emu g^{-1})	57.74	160.41	700.65	762.92	271.39

M$_s$ - Saturation magnetization
H$_c$ - Coercive field
M$_r$ - Remnant magnetization

2.5.5 La$_{1-x}$Sr$_x$MnO$_3$ manganites

The M-H loops recorded at room temperature for the La$_{1-x}$Sr$_x$MnO$_3$ (x=0.3, 0.4 and 0.5) samples are shown in figure 2.39. The magnetic parameters M$_s$, M$_r$ and H$_c$ for the La$_{1-x}$Sr$_x$MnO$_3$ (x=0.3, 0.4 and 0.5) samples are given in table 2.21.

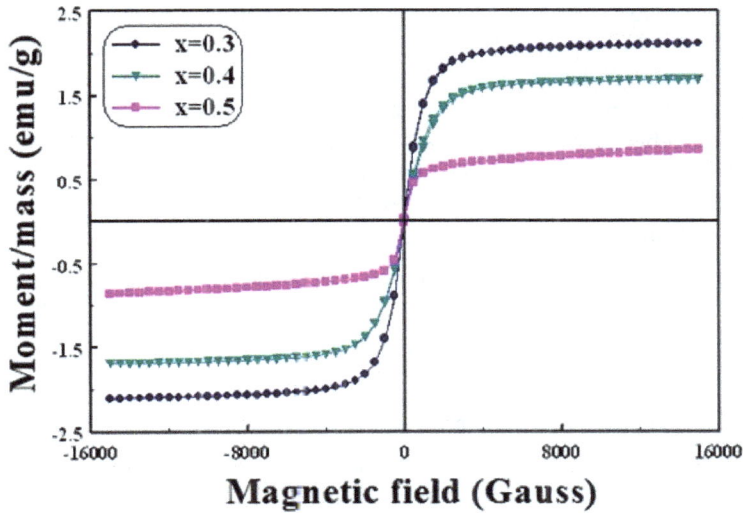

Figure 2.39 M-H curves for La$_{1-x}$Sr$_x$MnO$_3$, x=0.3, 0.4 and 0.5.

Table 2.21 Magnetic parameters for La$_{1-x}$Sr$_x$MnO$_3$, x=0.3, 0.4 and x=0.5.

Parameters	x=0.3	x=0.4	x=0.5
M$_s$ (emu g^{-1})	2.118	1.692	0.860
H$_c$ (G)	13.61	17.29	22.59
M$_r \times 10^{-3}$ (emu g^{-1})	24.62	19.62	22.00

M$_s$ - Saturation magnetization
H$_c$ - Coercive field
M$_r$ - Remnant magnetization

2.6 Charge density distribution studies by maximum entropy method

The structural refinement was carried out for the synthesized doped lanthanum chromite and lanthanum manganite materials using the Rietveld [Rietveld, 1969] procedure using the software JANA 2006 [Petříček et al., 2014]. The charge density distribution between the atoms in the lattice is analyzed by maximum entropy method [Collins, 1982] which uses the structure factors retrieved from the Rietveld [Rietveld, 1969] method. MEM method [Collins, 1982] gives precise pictures of distribution of charges in the valence region of the atoms. Hence, MEM technique [Collins, 1982] is helpful in analyzing the bonding features and other structure related parameters. In this work, this method has been executed by the software PRIMA (PRactice Iterative MEM Analysis) [Ruben and Fujio, 2004]. The resultant charge density distribution in the unit cell has been visualized through the visualization software VESTA (Visualization for Electronic and STructural Analysis) [Momma and Izumi, 2008].

In this section, the 3D, 2D and 1D electron density maps for all the doped lanthanum chromite and lanthanum manganite samples are presented. The bond lengths and mid bond electron densities for all prepared samples are tabulated and given in tables 2.22, 2.23, 2.24, 2.25, 2.26, 2.27, 2.28 and 2.29.

2.6.1 (Co, Mn) doped (La, Ca) based chromites - $(La_{0.8}Ca_{0.2})(Cr_{0.9-x}Co_{0.1}Mn_x)O_3$

Figures 2.40 (a) - (d) show the three dimensional electron density distributions in the unit cell of all the synthesized $(La_{0.8}Ca_{0.2})(Cr_{0.9-x}Co_{0.1}Mn_x)O_3$, (x=0.03, 0.06, 0.09 and 0.12) samples, with same iso-surface level of 3.0 e/Å^3. The 3D electron density distributions in La-O2 bond and Cr-O2 bond for the doped chromite samples are shown in figures 2.41 (a) - (d) and 2.41 (e) - (h) respectively.

Figures 2.42 (a) and 2.43 (a) illustrate the 3D unit cells with (101) plane and (020) plane shaded. The 2D electron density distribution between La and O2 atoms on the (101) plane is shown in figures 2.42 (b) - (e). The 2D electron density distribution between Cr and O2 atoms on (020) plane is shown in figures 2.43 (b) - (e).

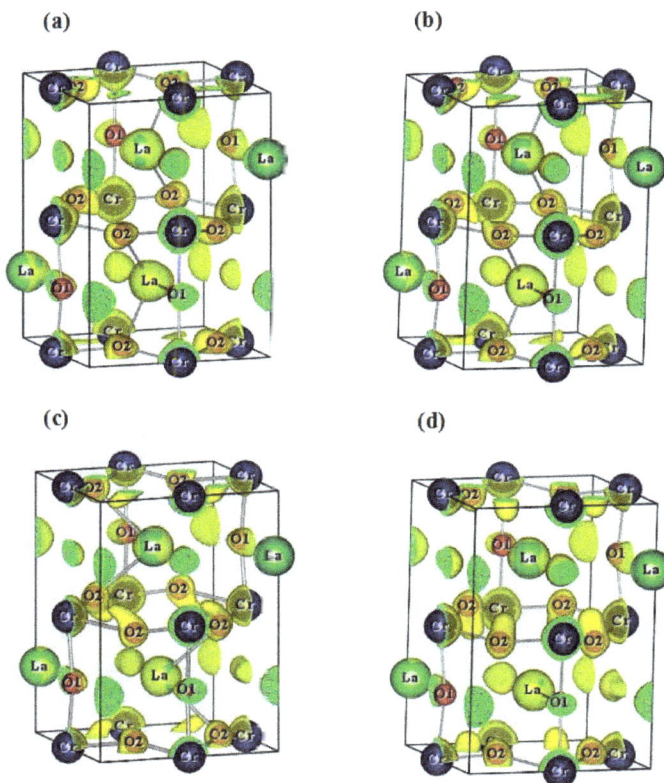

Figure 2.40 *Three dimensional electron density iso-surfaces for $(La_{0.8}Ca_{0.2})(Cr_{0.9-x}Co_{0.1}Mn_x)O_3$, (a) x=0.03, (b) x=0.05, (c) x=0.09 and (d) x=0.12 (iso-surface level: 3.0 e/$\AA^3$)*

La-O2   Cr-O2

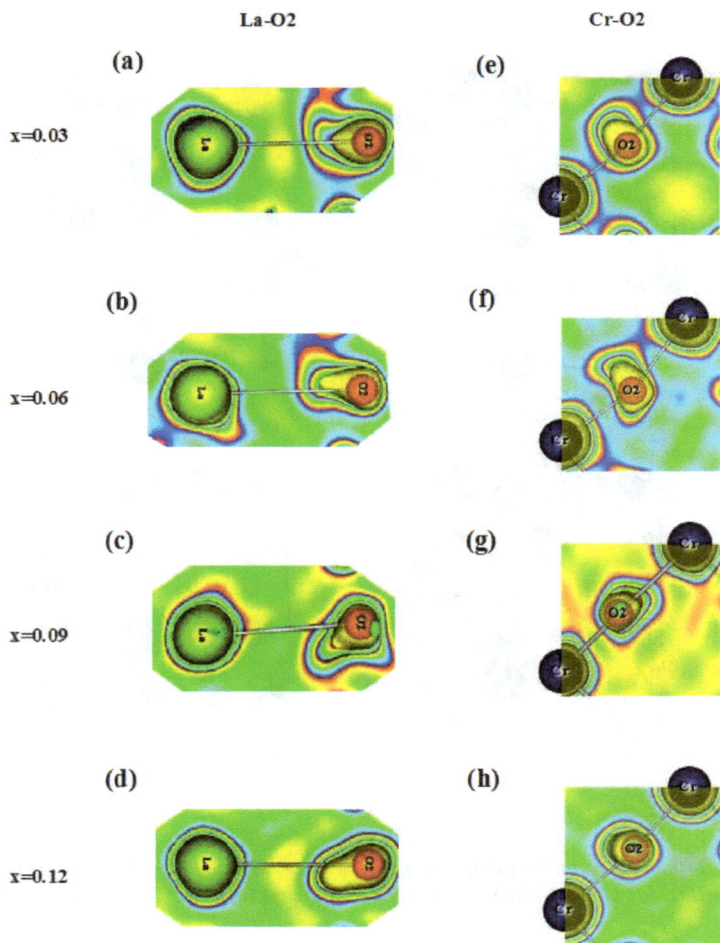

Figure 2.41 *3D electron density distributions for* $(La_{0.8}Ca_{0.2})(Cr_{0.9-x}Co_{0.1}Mn_x)O_3$ *along La-O2 bond **(a)** x=0.03, **(b)** x=0.06, **(c)** x=0.09 and **(d)** x=0.12 and along Cr-O2 bond **(e)** x=0.03, **(f)** x=0.06, **(g)** x=0.09 and **(h)** x=0.12.*

(a)

(b)

(c)

(d)

(e)

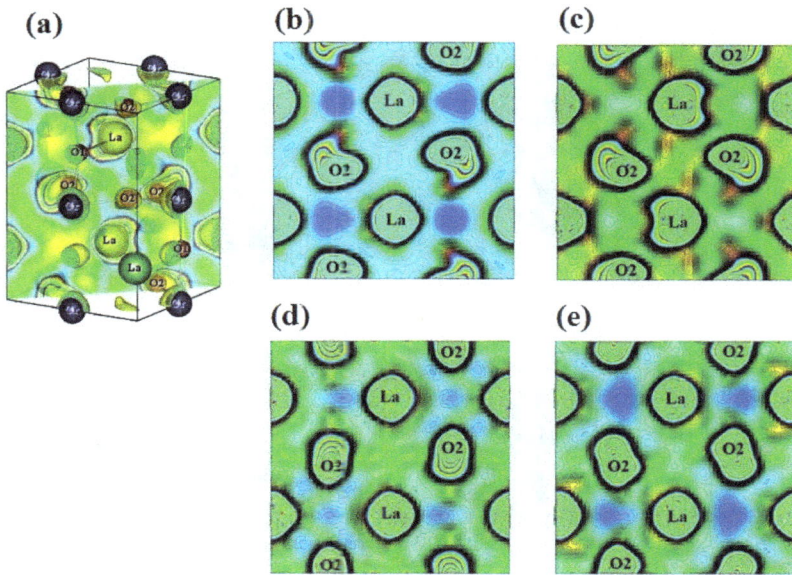

Figure 2.42 (a) *3D unit cell of $(La_{0.8}Ca_{0.2})(Cr_{0.87}Co_{0.1}Mn_{0.03})O_3$ with (101) plane shaded. Two dimensional electron density distribution on (101) plane for $(La_{0.8}Ca_{0.2})(Cr_{0.9-x}Co_{0.1}Mn_x)O_3$, **(b)** x =0.03, **(c)** x=0.06 **(d)** x=0.09 and **(e)** x=0.12 (contour range: 0-1.0 $e/Å^3$, contour interval: 0.04 $e/Å^3$).*

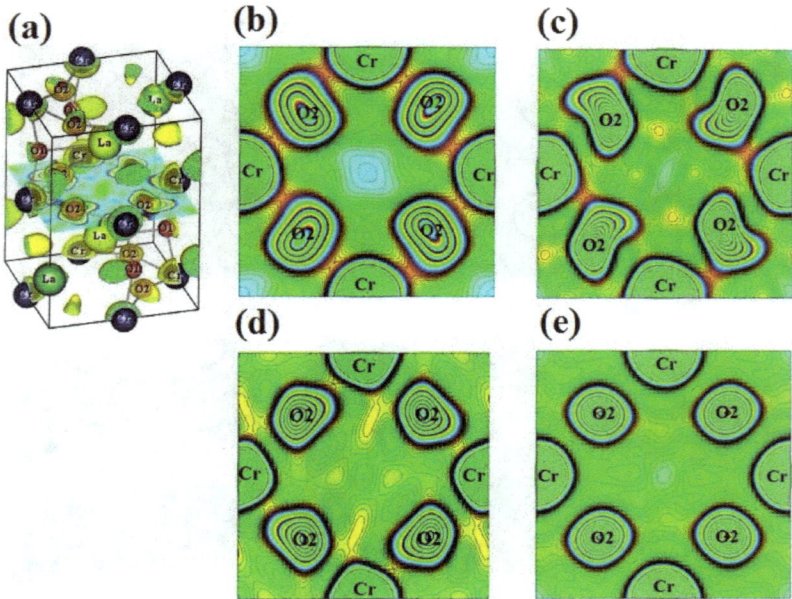

Figure 2.43 (a) *3D unit cell of $(La_{0.8}Ca_{0.2})(Cr_{0.87}Co_{0.1}Mn_{0.03})O_3$ with (020) plane shaded. Two dimensional electron density distribution on (020) plane for $(La_{0.8}Ca_{0.2})(Cr_{0.9-x}Co_{0.1}Mn_x)O_3$, (b) x=0.03, (c) x=0.06, (d) x=0.09 and (e) x=0.12 (contour range: 0-1.0 $e/Å^3$, contour interval: 0.04 $e/Å^3$).*

The one dimensional electron density profiles for the bonds La-O2 and Cr-O2 are shown in figures 2.44 and 2.45. The bond length and the mid bond density values of La-O2 and Cr-O2 bonds are presented in table 2.22. The oxygen bonds O1-O2 and O2(A)-O2(B) in the 3D unit cell of $(La_{0.8}Ca_{0.2})(Cr_{0.87}Co_{0.1}Mn_{0.03})O_3$ are shown in figure 2.46. Figures 2.47 and 2.48 show the one dimensional electron density profiles for oxygen bonds O1-O2 and O2(A)-O2(B). Table 2.23 gives the bond lengths and mid bond density values for O1-O2 and O2(A)-O2(B) bonds.

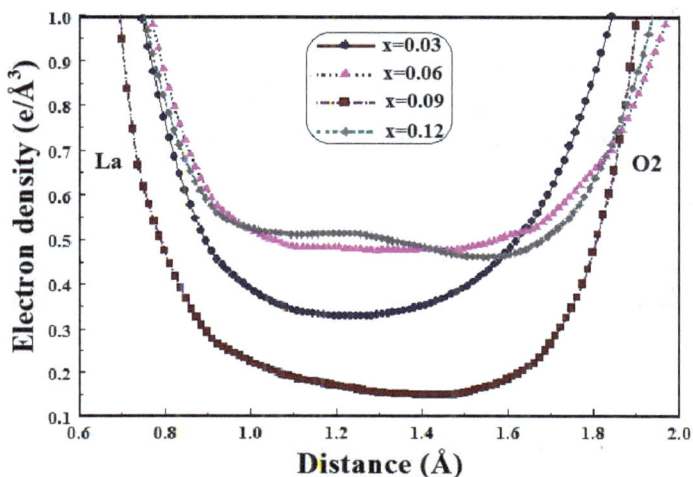

Figure 2.44 One dimensional electron density profiles along La and O2 atoms in $(La_{0.8}Ca_{0.2})(Cr_{0.9-x}Co_{0.1}Mn_x)O_3$, $x=0.03$, 0.06, 0.09, 0.12.

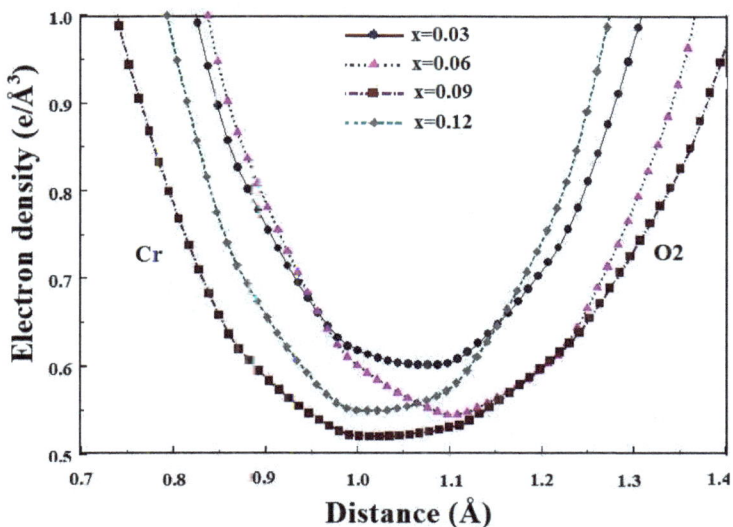

Figure 2.45 One dimensional electron density profiles along Cr and O2 atoms in $(La_{0.8}Ca_{0.2})(Cr_{0.9-x}Co_{0.1}Mn_x)O_3$, $x=0.03$, 0.06, 0.09, 0.12.

Table 2.22 *Bond lengths and mid bond electron densities for La- O2 & Cr-O2 bonds for* $(La_{0.8}Ca_{0.2})(Cr_{0.9-x}Co_{0.1}Mn_x)O_3$, *x=0.03, 0.06, 0.09, 0.12.*

| Samples | Bonding | | | |
| | La-O2 | | Cr-O2 | |
	Bond length (Å)	Mid bond electron density (e/Å³)	Bond length (Å)	Mid bond electron density (e/Å³)
x=0.03	2.478	0.328	2.164	0.601
x=0.06	2.476	0.475	2.163	0.545
x=0.09	2.485	0.147	2.168	0.520
x=0.12	2.474	0.461	2.161	0.550

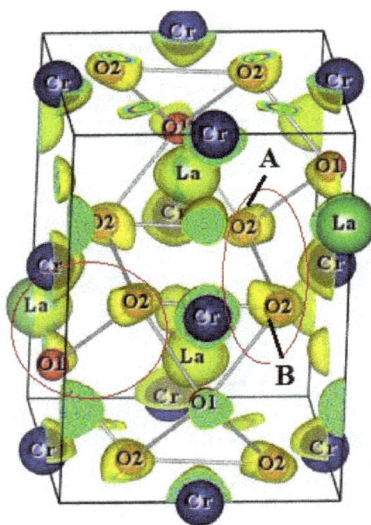

Figure 2.46 *The oxygen bonds O1-O2 and O2(A)-O2(B) in the three dimensional unit cell of* $(La_{0.8}Ca_{0.2})(Cr_{0.87}Co_{0.1}Mn_{0.03})O_3$.

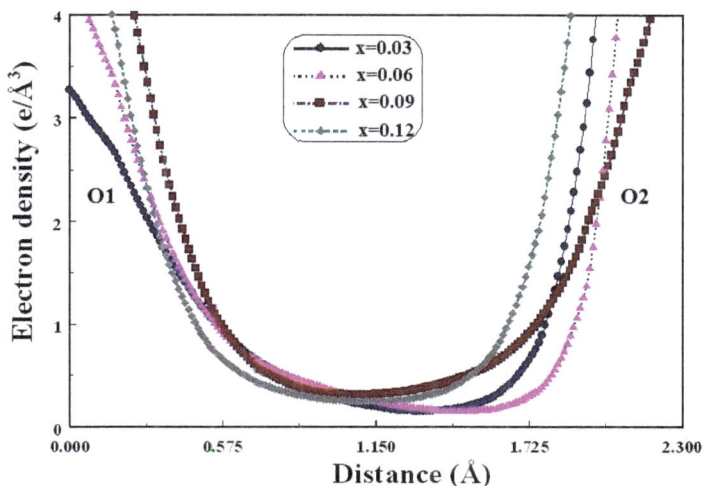

Figure 2.47 One dimensional electron density profiles along O1 and O2 atoms in $(La_{0.8}Ca_{0.2})(Cr_{0.9-x}Co_{0.1}Mn_x)O_3$, $x=0.03$, 0.06, 0.09, 0.12.

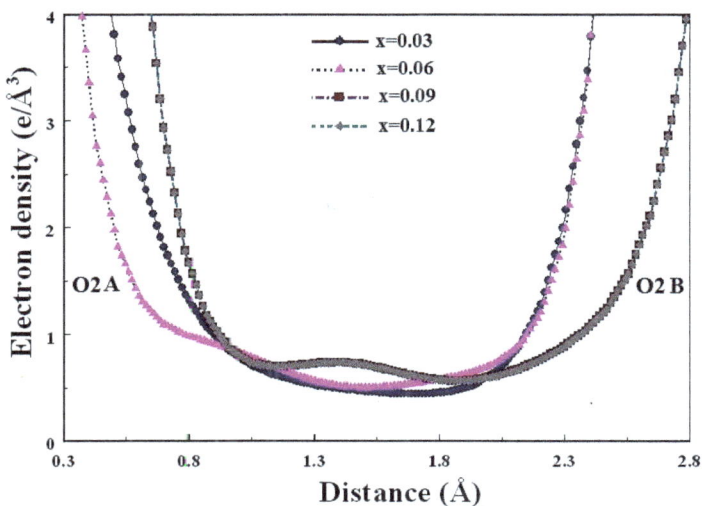

Figure 2.48 One dimensional electron density profiles along O2(A) and O2(B) atoms in $(La_{0.8}Ca_{0.2})(Cr_{0.9-x}Co_{0.1}Mn_x)O_3$, $x=0.03$, 0.06, 0.09, 0.12.

Table 2.23 Bond lengths and mid bond electron densities of O1- O2 & O2(A)-O2(B) bonds for $(La_{0.8}Ca_{0.2})(Cr_{0.9-x}Co_{0.1}Mn_x)O_3$, x=0.03, 0.06, 0.09, 0.12.

Samples	Bonding			
	O1-O2		O2(A)-O2(B)	
	Bond length (Å)	Mid bond electron density (e/Å³)	Bond length (Å)	Mid bond electron density (e/Å³)
x=0.03	2.401	0.154	2.850	0.444
x=0.06	2.399	0.148	2.849	0.505
x=0.09	2.410	0.305	2.841	0.567
x=0.12	2.397	0.244	2.843	0.480

2.6.2 (Co, Fe) doped (La, Ca) based chromites - $(La_{0.8}Ca_{0.2})(Cr_{0.9-x}Co_{0.1}Fe_x)O_3$

The 3D electron density distributions in the unit cell with similar iso-surface level of 3.0 $e/Å^3$ for the synthesized $(La_{0.8}Ca_{0.2})(Cr_{0.9-x}Co_{0.1}Fe_x)O_3$, (x=0.03, 0.06, 0.09 and 0.12) samples are shown in figures 2.49 (a) - (d). The 3D electron density distribution for the La-O2 and Cr-O2 bonds for all Fe concentrations (x=0.03, 0.06, 0.09 and 0.12) are shown in figures 2.50 (a) - (d) and 2.50 (e) - (h) respectively.

Figures 2.51 (a) and 2.52 (a) show the three dimensional unit cells with the (101) plane and (020) plane shaded. The 2D electron density distributions for La-O2 and Cr-O2 bonds are drawn in the range 0-1.0 $e/Å^3$ with an interval 0.04 $e/Å^3$ on the (101) plane and (020) plane and are shown in figures 2.51 (b) - (e) and 2.52 (b) - (e) respectively.

The one dimensional electron density profiles for the bonds La-O2, Cr-O2 and O1-O2 are drawn and shown in figures 2.53, 2.54 and 2.55 respectively. The bond lengths and mid bond density values for La-O2, Cr-O2 and O1-O2 are tabulated in table 2.24.

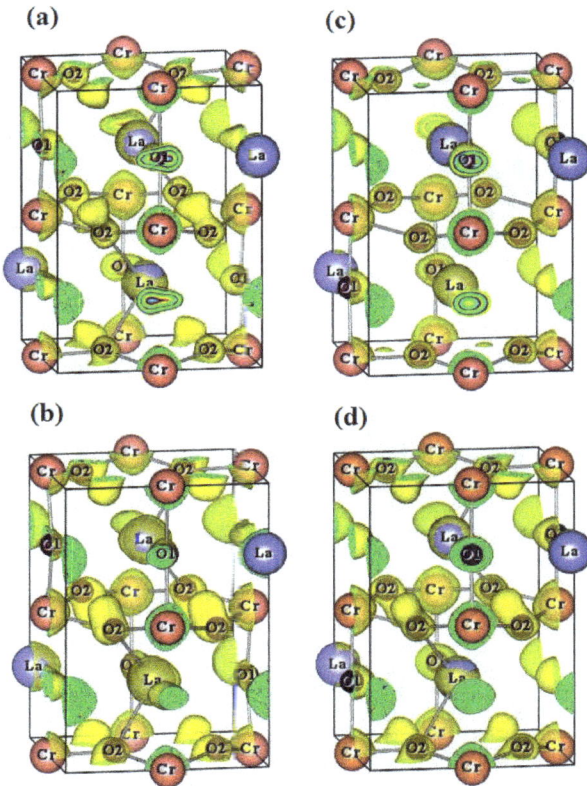

Figure 2.49 *Three dimensional electron density iso-surfaces for* $(La_{0.8}Ca_{0.2})(Cr_{0.9-x}Co_{0.1}Fe_x)O_3$, *(a)* $x=0.03$, *(b)* $x=0.06$, *(c)* $x=0.09$ *and (d)* $x=0.12$ *(iso-surface level: 3.0 $e/Å^3$).*

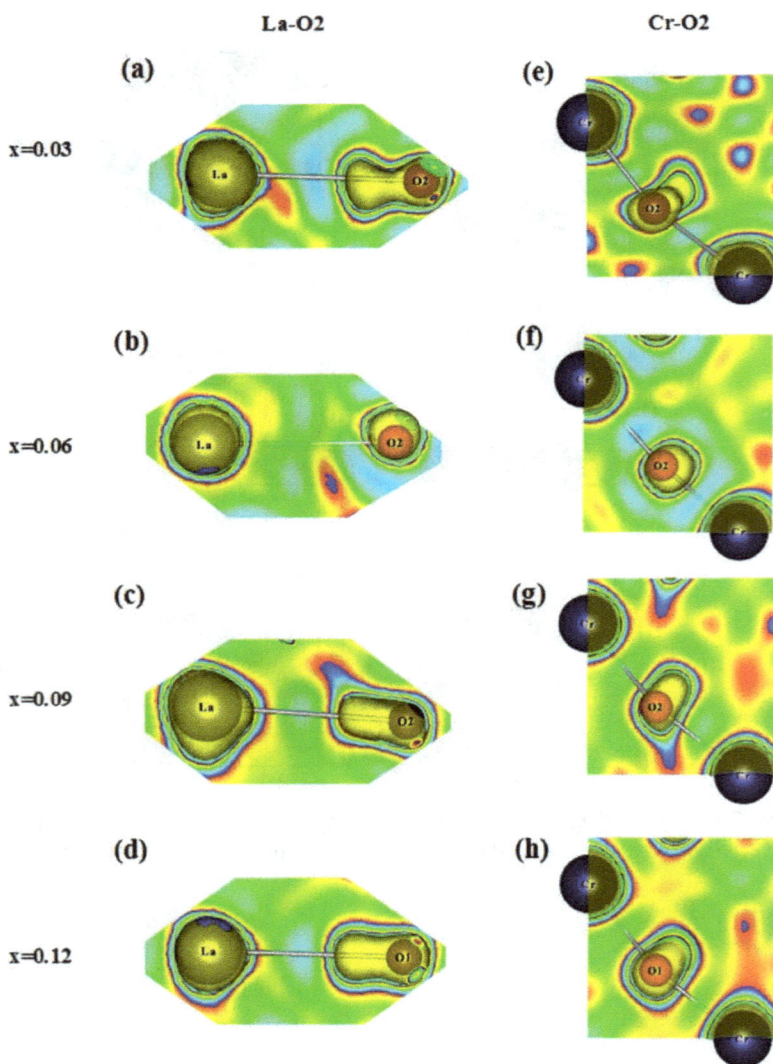

Figure 2.50 *3D electron density distributions for* $(La_{0.8}Ca_{0.2})(Cr_{0.9-x}Co_{0.1}Fe_x)O_3$ *along La-O2 bond* **(a)** *x=0.03,* **(b)** *x=0.06,* **(c)** *x=0.09 and* **(d)** *x=0.12 and along Cr-O2 bond* **(e)** *x=0.03,* **(f)** *x=0.06,* **(g)** *x=0.09 and* **(h)** *x=0.12.*

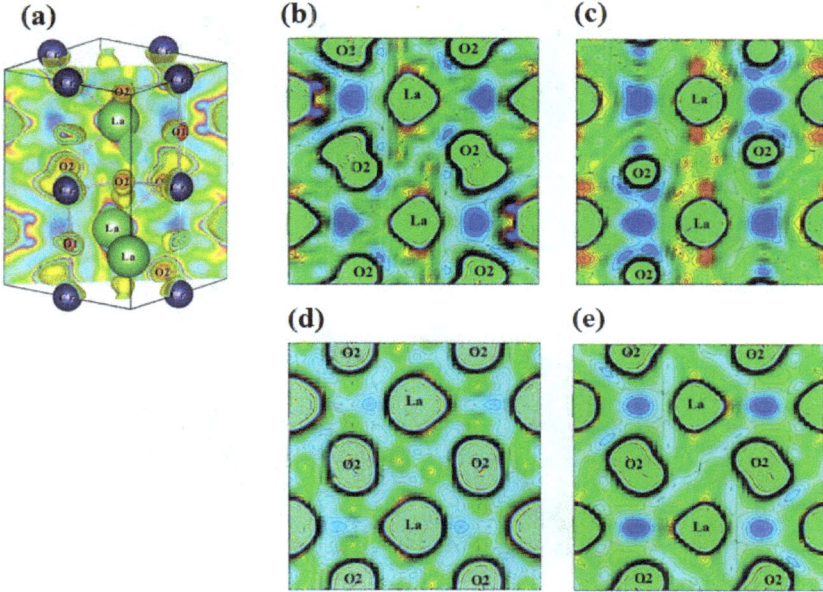

Figure 2.51 (a) *3D unit cell of $(La_{0.8}Ca_{0.2})(Cr_{0.87}Co_{0.1}Fe_{0.03})O_3$ with (101) plane shaded. Two dimensional electron density distribution on the (101) plane for $(La_{0.8}Ca_{0.2})(Cr_{0.9-x}Co_{0.1}Fe_x)O_3$, **(b)** x=0.03, **(c)** x=0.06, **(d)** x=0.09 and **(e)** x=0.12 (contour range: 0-1.0 $e/Å^3$, contour interval: 0.04 $e/Å^3$).*

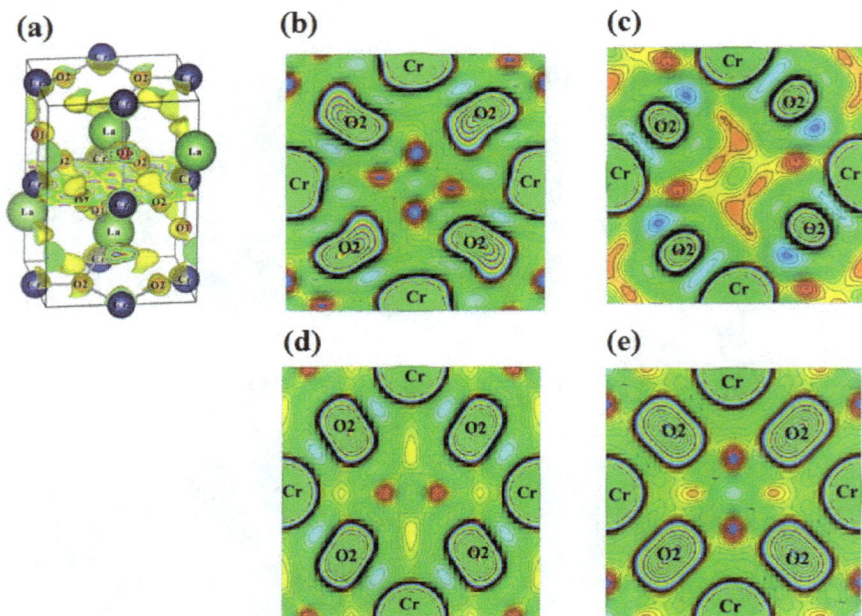

Figure 2.52 (a) *3D unit cell of $(La_{0.8}Ca_{0.2})(Cr_{0.87}Co_{0.1}Fe_{0.03})O_3$ with (020) plane shaded. Two dimensional electron density distribution on (020) plane for $(La_{0.8}Ca_{0.2})(Cr_{0.9-x}Co_{0.1}Fe_x)O_3$,* **(b)** *x=0.03,* **(c)** *x=0.06,* **(d)** *x=0.09 and* **(e)** *x=0.12 (contour range: 0-1.0 $e/Å^3$, contour interval: 0.04 $e/Å^3$).*

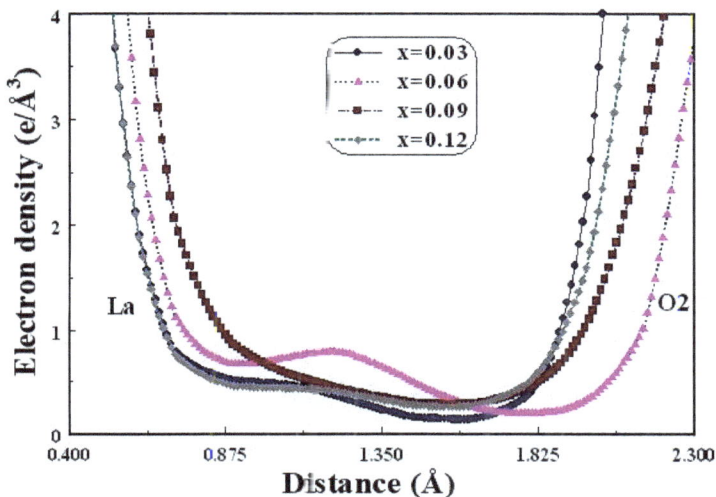

Figure 2.53 *One dimensional electron density profiles along La and O2 atoms in $(La_{0.8}Ca_{0.2})(Cr_{0.9-x}Co_{0.1}Fe_x)O_3$, x=0.03, 0.06, 0.09, 0.12.*

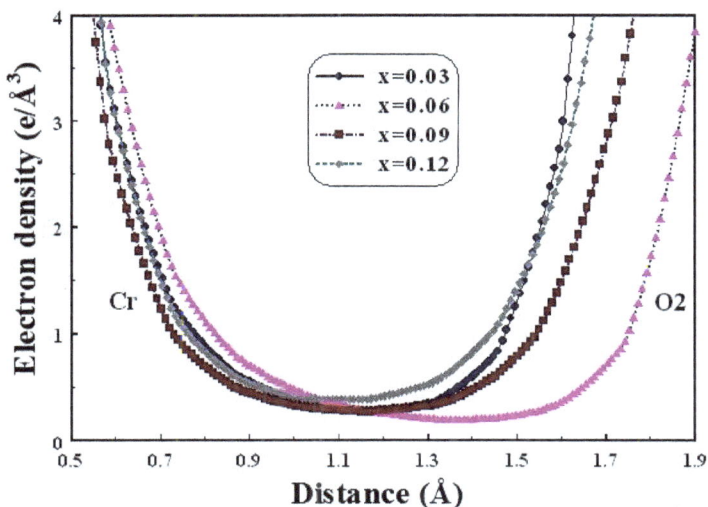

Figure 2.54 *One dimensional electron density profiles along Cr and O2 atoms in $(La_{0.8}Ca_{0.2})(Cr_{0.9-x}Co_{0.1}Fe_x)O_3$ x=0.03, 0.06, 0.09, 0.12.*

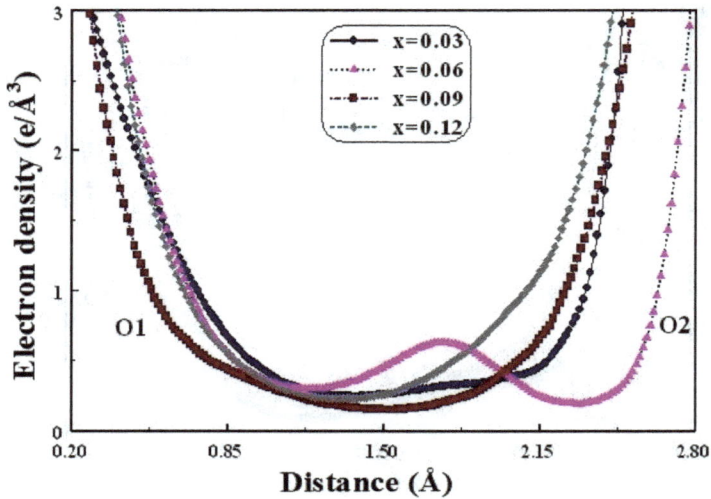

Figure 2.55 One dimensional electron density profiles along O1 and O2 atoms in $(La_{0.8}Ca_{0.2})(Cr_{0.9-x}Co_{0.1}Fe_x)O_3$, x=0.03, 0.06, 0.09, 0.12.

Table 2.24 Bond lengths and mid bond electron densities of La-O2, Cr-O2 and O1-O2 bonds for $(La_{0.8}Ca_{0.2})(Cr_{0.9-x}Co_{0.1}Fe_x)O_3$, x=0.03, 0.06, 0.09, 0.12.

Samples	Bonding					
	La-O2		Cr-O2		O1-O2	
	Bond length (Å)	Mid bond electron density (e/Å³)	Bond length (Å)	Mid bond electron density (e/Å³)	Bond length (Å)	Mid bond electron density (e/Å³)
x=0.03	2.395	0.368	1.947	0.421	2.885	0.248
x=0.06	2.398	0.802	1.949	0.522	2.887	0.406
x=0.09	2.388	0.455	1.940	0.357	2.874	0.152
x=0.12	2.392	0.409	1.946	0.441	2.880	0.239

2.6.3 (Co, Cu) doped (La, Ca) based chromites - $(La_{0.8}Ca_{0.2})(Cr_{0.9-x}Co_{0.1}Cu_x)O_3$

The 3D electron density distributions for $(La_{0.8}Ca_{0.2})(Cr_{0.9-x}Co_{0.1}Cu_x)O_3$, (x=0.00, 0.03 and 0.12) samples are constructed with an iso-surface level of 3.0 e/$Å^3$ and are presented in figures 2.56 (a) - (c). Figure 2.57 (a) illustrates the 3D unit cell with (101) plane shaded. Figures 2.57 (b) - (d) show the 2D electron density contour maps corresponding to (101) plane for $(La_{0.8}Ca_{0.2})(Cr_{0.9-x}Co_{0.1}Cu_x)O_3$ (x=0.00, 0.03 and 0.12) samples. Figure 2.58 (a) represents the 3D unit cell with (020) plane shaded and figures 2.58 (b) - (d) show the 2D electron density contour maps corresponding to (020) plane with Cr-O2 bonding.

The 1D electron density profiles for La-O2 and Cr-O2 bonds are shown in figures 2.59 and 2.60 respectively. Figure 2.61 shows the 1D profile for oxygen bond O1-O2. Table 2.25 gives the bond lengths and the mid bond density values for La-O2, Cr-O2 and O1-O2 bonds. The CrO_6 octahedron constructed using VESTA [Momma and Izumi, 2008] for the sample $(La_{0.8}Ca_{0.2})(Cr_{0.87}Co_{0.1}Cu_{0.03})O_3$ is shown in figure 2.62. Table 2.26 gives the bond lengths for O1-O1, O2(2)-O2(4) and O2(1)-O2(3) bonds for all the Cu concentrations of the synthesized samples.

(a)

(b)

(c)

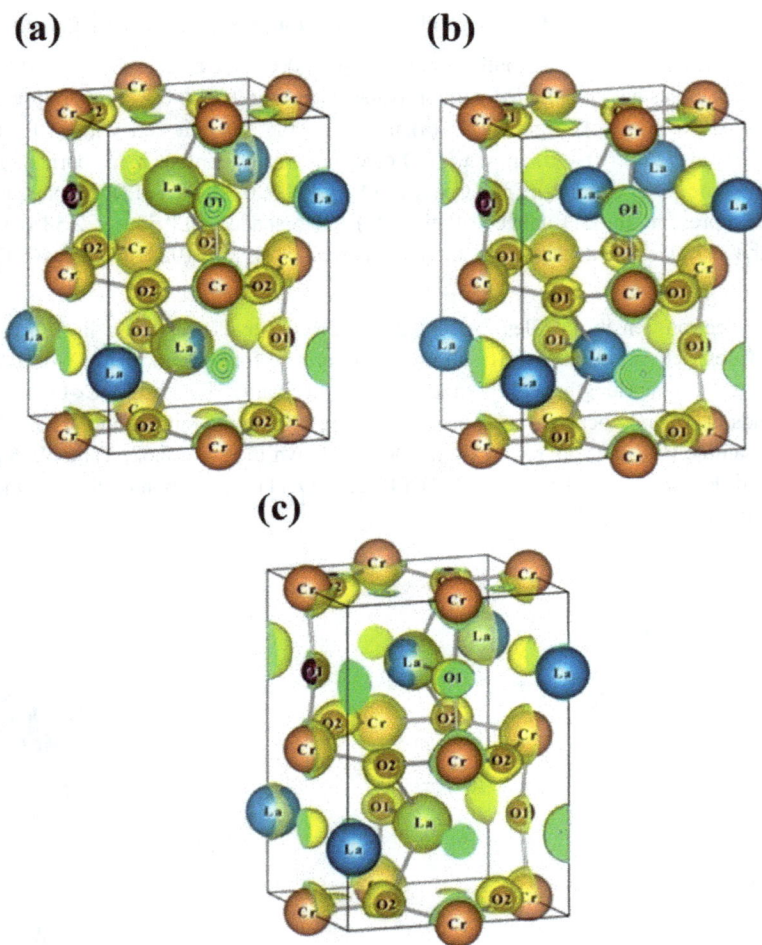

Figure 2.56 *Three dimensional electron density iso-surfaces for* $(La_{0.8}Ca_{0.2})(Cr_{0.9-x}Co_{0.1}Cu_x)O_3$, *(a)* $x=0.00$, *(b)* $x=0.03$ *and* *(c)* $x=0.12$, *(iso- surface level: 3.0 $e/Å^3$).*

(a)

(b)

(c)

(d)

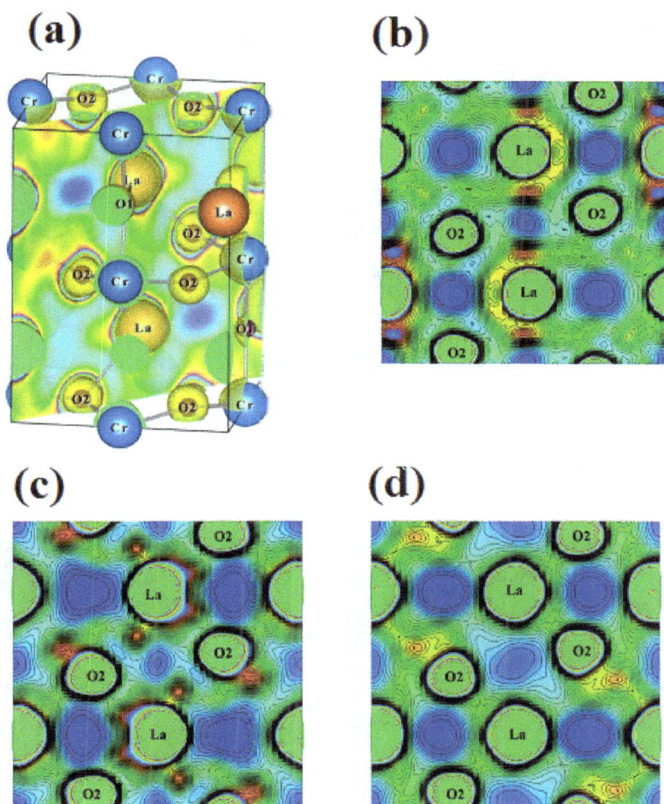

Figure 2.57 (a) *3D unit cell of* $(Lc_{0.8}Ca_{0.2})(Cr_{0.87}Co_{0.1}Cu_{0.03})O_3$ *with (101) plane shaded. Two dimensional electron density distribution on (101) plane for* $(La_{0.8}Ca_{0.2})(Cr_{0.9-x}Co_{0.1}Cu_x)O_3$, *(b)* $x=0.00$, *(c)* $x=0.03$ *and (d)* $x=0.12$ *(contour range: 0-1.0 e/$Å^3$, contour interval: 0.04 e/$Å^3$).*

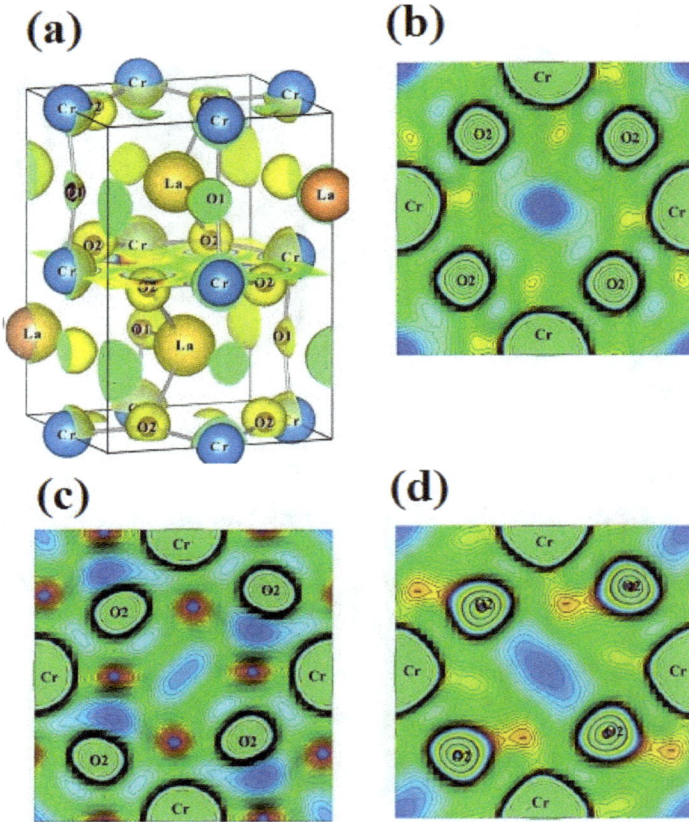

Figure 2.58 (a) *3D unit cell of $(La_{0.8}Ca_{0.2})(Cr_{0.87}Co_{0.1}Cu_{0.03})O_3$ with (020) plane shaded. Two dimensional electron density distribution on (020) plane for $(La_{0.8}Ca_{0.2})(Cr_{0.9-x}Co_{0.1}Cu_x)O_3$, **(b)** x=0.00, **(c)** x=0.03 and **(d)** x=0.12 (contour range: 0-1.0 e/$\AA^3$, contour interval: 0.04 e/$\AA^3$).*

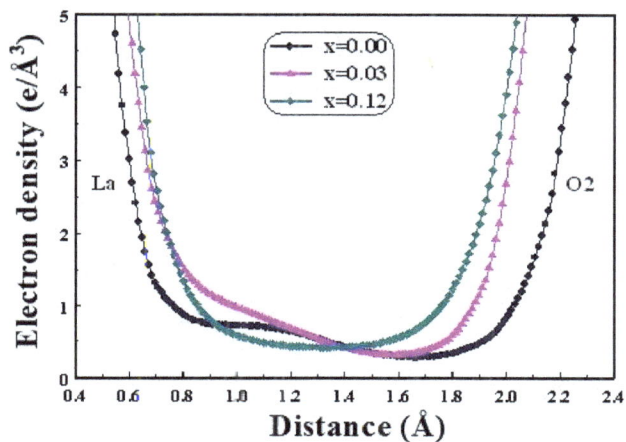

Figure 2.59 One dimensional electron density profiles along La and O2 atoms in $(La_{0.8}Ca_{0.2})(Cr_{0.9-x}Co_{0.1}Cu_x)O_3$, $x=0.00$, 0.03, 0.12.

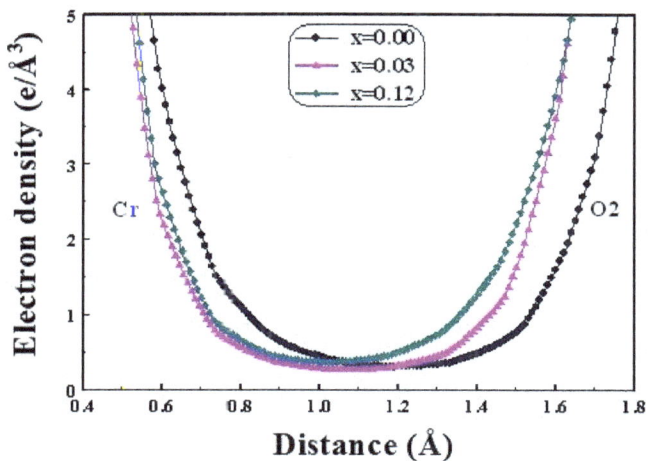

Figure 2.60 One dimensional electron density profiles along Cr and O2 atoms in $(La_{0.8}Ca_{0.2})(Cr_{0.9-x}Co_{0.1}Cu_x)O_3$, $x=0.00$, 0.03, 0.12.

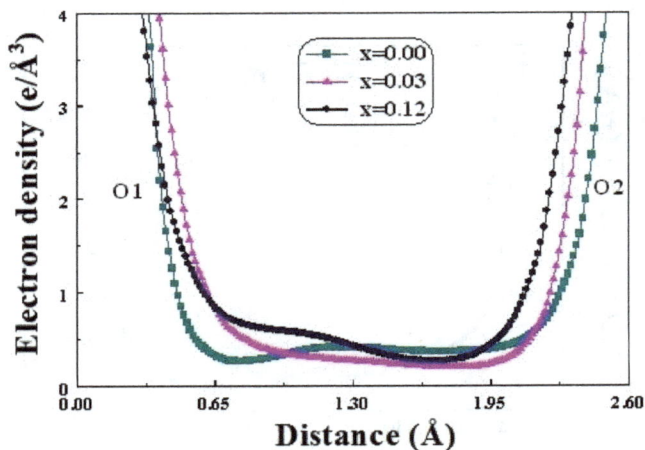

Figure 2.61 *One dimensional electron density profiles along O1 and O2 atoms in* $(La_{0.8}Ca_{0.2})(Cr_{0.9-x}Co_{0.1}Cu_x)O_3$, *x=0.00, 0.03, 0.12.*

Table 2.25 *Bond lengths and mid bond electron densities of La-O2, Cr-O2 and O1-O2 bonds for* $(La_{0.8}Ca_{0.2})(Cr_{0.9-x}Co_{0.1}Cu_x)O_3$, *x=0.00, 0.03, 0.12.*

	Bonding					
	La-O2		Cr-O2		O1-O2	
Samples	Bond length (Å)	Mid bond electron density (e/Å³)	Bond length (Å)	Mid bond electron density (e/Å³)	Bond length (Å)	Mid bond electron density (e/Å³)
x=0.00	2.481	0.624	1.990	0.504	2.791	0.401
x=0.03	2.469	0.385	1.977	0.391	2.776	0.259
x=0.12	2.454	0.303	1.966	0.429	2.757	0.381

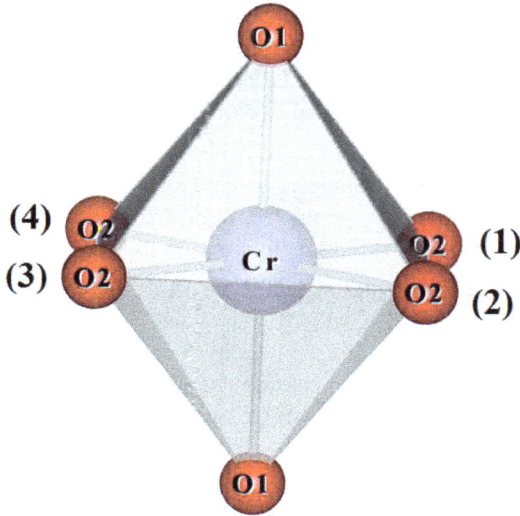

Figure 2.62 The CrO_6 octahedron of $(La_{0.8}Ca_{0.2})$ $(Cr_{0.87}Co_{0.1}Cu_{0.03})$ O_3 obtained from VESTA.

Table 2.26 Bond lengths for (O1-O1, O2(2)-O2(4) and O2(1)-O2(3) bonds of CrO_6 octahedron of $(La_{0.8}Ca_{0.2})(Cr_{0.9-x}Co_{1}Cu_x)O_3$, x=0.00, 0.03, 0.12.

Bond	Bond length (Å)		
	x=0.00	x=0.03	x=0.12
O1-O1	3.971	3.932	3.910
O2(2)-O2(4)	3.754	3.806	3.937
O2(1)-O2(3)	4.162	4.075	3.970

2.6.4 La$_{1-x}$Ca$_x$MnO$_3$ manganites

Three dimensional electron density distributions in the unit cell for calcium doped lanthanum manganite samples with similar iso-surface level of 3.0 e/Å^3 are shown in figures 2.63 (a) - (e). Figures 2.64 (a) and 2.65 (a) present the 3D unit cells with the (101) plane and (020) plane shaded.

The 2D electron density distribution maps for the La-O2 bond are drawn in the range 0-1.0 e/Å^3 with an interval 0.04 e/Å^3 on the (101) plane (figures 2.64 (b) - (f)) and for Mn-O2 bond in the same range in (020) plane (figures 2.65 (b) - (f)).

Figures 2.66 and 2.67 (a) - (c) illustrate the 1D electron density profiles for La-O2 and Mn-O bonds (Mn-O2 along a-axis, Mn-O1 along b-axis and Mn-O2 along c-axis). Table 2.27 gives the bond length and mid bond density values for the La-O2 bond. Table 2.28 gives the bond lengths and mid bond electron densities for all Mn-O bonds (Mn-O2 (a-axis), Mn-O1 (b-axis) and Mn-O2 (c-axis)) for La$_{1-x}$Ca$_x$MnO$_3$ (x=0.1, 0.2, 0.3, 0.4, 0.5).

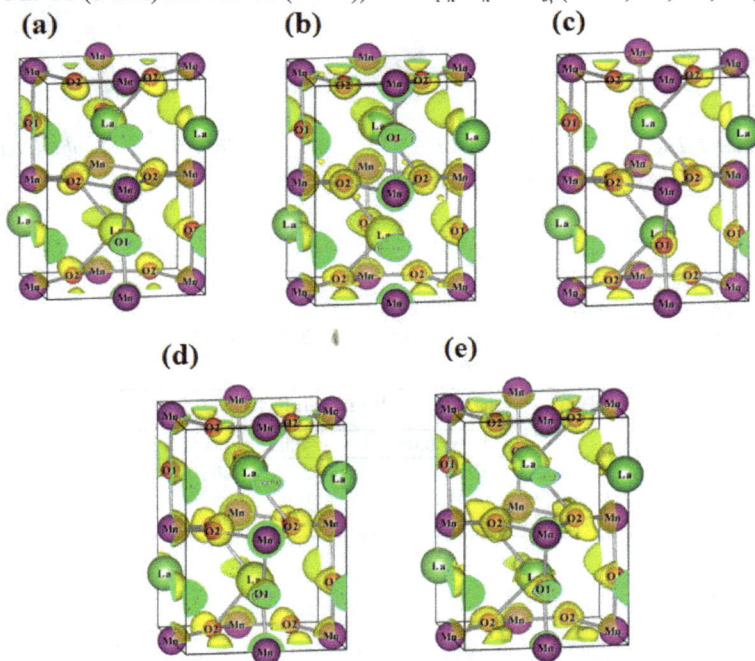

Figure 2.63 Three dimensional electron density iso-surfaces for La$_{1-x}$Ca$_x$MnO$_3$, *(a)* x=0.1, *(b)* x=0.2, *(c)* x=0.3, *(d)* x=0.4 and *(e)* x=0.5 (iso-surface level: 3.0 e/Å^3).

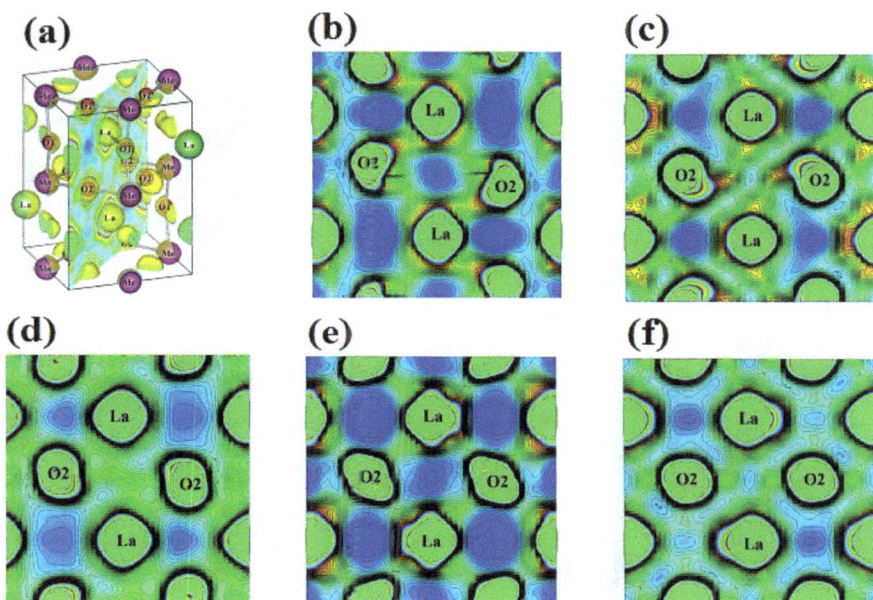

Figure 2.64 (a) *3D unit cell of La$_{1-x}$Ca$_x$MnO$_3$with (101) plane shaded. Two dimensional electron density distribution on (101) plane for La$_{1-x}$Ca$_x$MnO$_3$,* **(b)** *x=0.1,* **(c)** *x=0.2,* **(d)** *x=0.3,* **(e)** *x=0.4 and* **(f)** *x=0.5 (contour range: 0-1.0 e/Å^3, contour interval: 0.04 e/Å^3).*

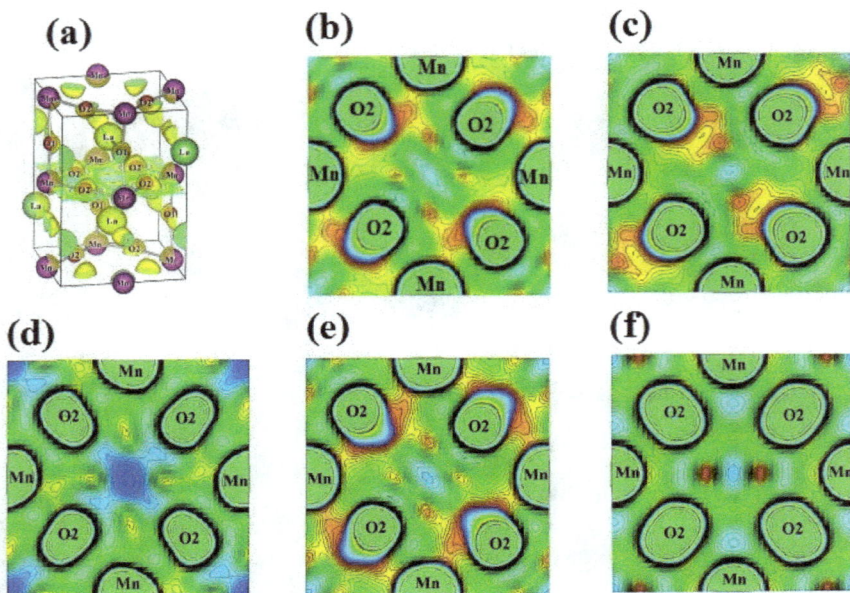

Figure 2.65 *(a)* *3D unit cell of $La_{1-x}Ca_xMnO_3$ with the (020) plane shaded. Two dimensional electron density distribution on the (020) plane for $La_{1-x}Ca_xMnO_3$, (b) x=0.1, (c) x=0.2, (d) x=0.3, (e) x=0.4 and (f) x=0.5 (contour range: 0-1.0 $e/Å^3$, contour interval: 0.04 $e/Å^3$).*

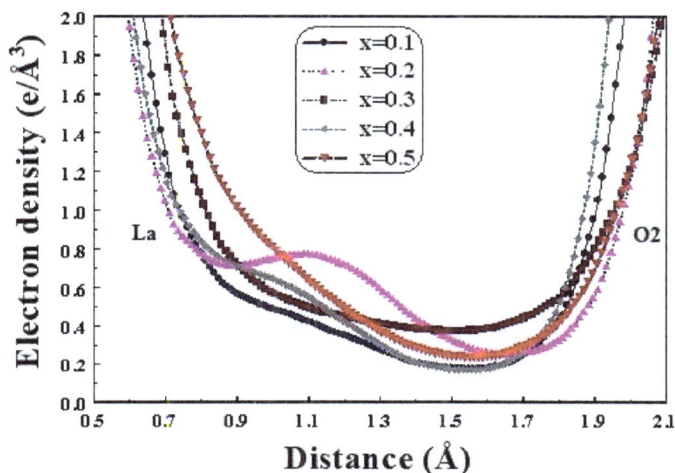

Figure 2.66 *One dimensional electron density profiles along the La and O2 atoms in La$_{1-x}$Ca$_x$MnO$_3$, x=0.1, 0.2, 0.3, 0.4, 0.5*

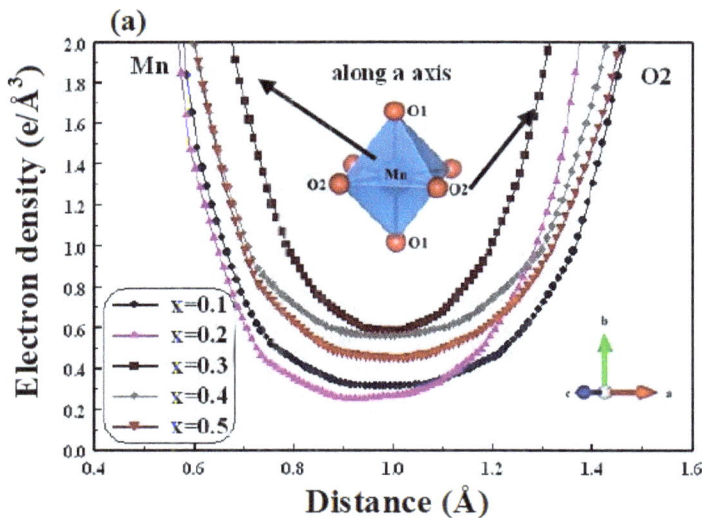

Figure 2.67 (a) *One dimensional electron density profiles along Mn and O2 along the a-axis in La$_{1-x}$Ca$_x$MnO$_3$, x=0.1, 0.2, 0.3, 0.4, 0.5.*

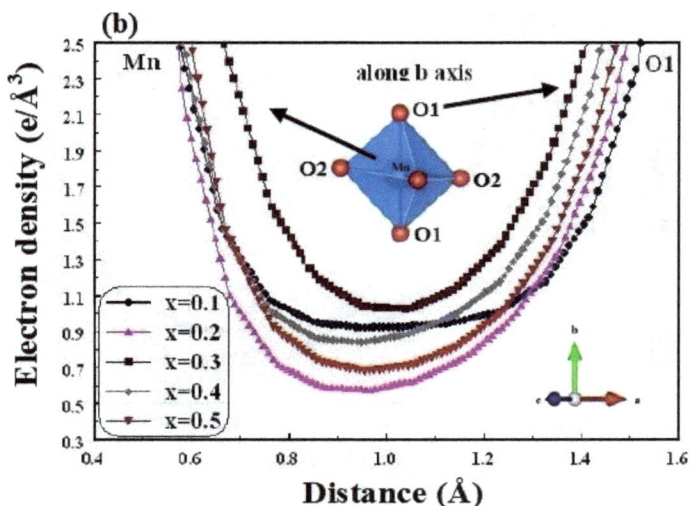

Figure 2.67 (b) *One dimensional electron density profiles along Mn and O1 along the b-axis in La$_{1-x}$Ca$_x$MnO$_3$, x=0.1, 0.2, 0.3, 0.4, 0.5.*

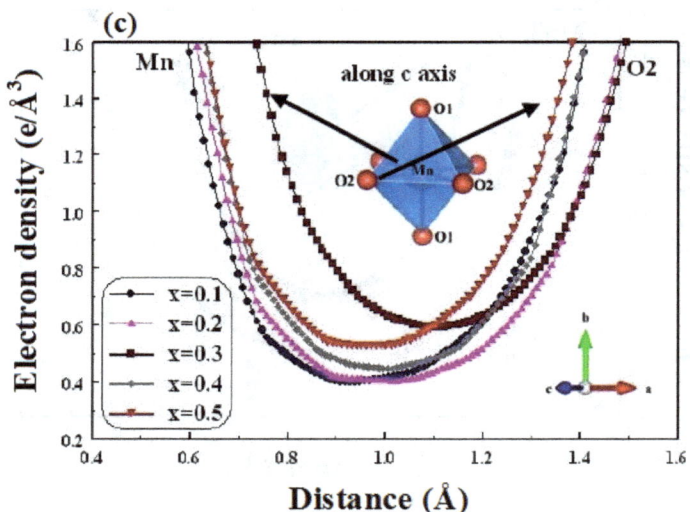

Figure 2.67 (c) *One dimensional electron density profiles along Mn and O2 along the c-axis in La$_{1-x}$Ca$_x$MnO$_3$, x=0.1, 0.2, 0.3, 0.4, 0.5.*

Table 2.27 *Bond lengths and mid bond electron densities of the La-O2 bond for* $La_{1-x}Ca_xMnO_3$ $x=0.1, 0.2, 0.3, 0.4, 0.5.$

Samples	La-O2	
	Bond length (Å)	Mid bond electron density (e/Å^3)
x=0.1	2.484	0.318
x=0.2	2.482	0.667
x=0.3	2.471	0.436
x=0.4	2.458	0.397
x=0.5	2.448	0.452

Table 2.28 *Bond lengths and mid bond electron densities of the Mn-O2 (a-axis), the Mn-O1 (b-axis) and the Mn-O2 (c-axis) bonds for* $La_{1-x}Ca_xMnO_3$ $x=0.1, 0.2, 0.3, 0.4, 0.5.$

Samples	Mn-O2 (a-axis)		Mn-O1 (b-axis)		Mn-O2 (c-axis)	
	Bond length (Å)	Mid bond electron density (e/Å^3)	Bond length (Å)	Mid bond electron density (e/Å^3)	Bond length (Å)	Mid bond electron density (e/Å^3)
x=0.1	1.987	0.268	1.989	0.992	1.961	0.389
x=0.2	1.973	0.354	1.987	0.784	1.958	0.588
x=0.3	1.969	0.468	1.982	0.514	1.953	0.489
x=0.4	1.958	0.539	1.968	1.025	1.942	0.676
x=0.5	1.953	0.423	1.960	0.589	1.935	0.531

2.6.5 $La_{1-x}Sr_xMnO_3$ manganites

The 3D charge density distributions of the synthesized $La_{1-x}Sr_xMnO_3$ (x=0.3, 0.4 and 0.5) manganite samples are constructed with an iso-surface level of 1.5 e/Å^3 and are shown in figures 2.68 (a) - (c). Figures 2.69 (a) and figure 2.70 (a) show the three dimensional unit cells with the (004) and (012) planes shaded. The 2D maps for the La-O bond in the (004) plane and the Mn-O bond on the (012) plane are shown in figures 2.69 (b) – (d) and 2.70 (b) – (d) respectively. The one dimensional profiles for the La-O, Mn-O and O-O bonds are shown in figures 2.71, 2.72 and 2.73 respectively. The bond lengths and mid bond electron density values for the La-O, Mn-O and O-O bonds are tabulated in table 2.29.

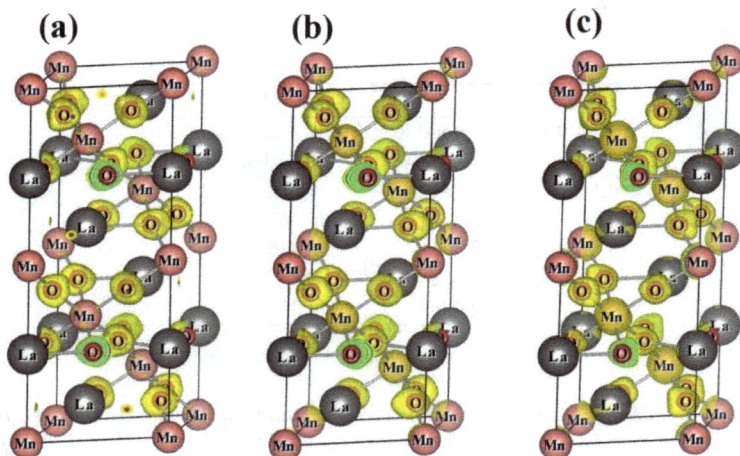

Figure 2.68 Three dimensional electron density iso-surfaces for of $La_{1-x}Sr_xMnO_3$, **(a)** x=0.3, **(b)** x= 0.4 and **(c)** x=0.5 (iso-surface level: 1.5 $e/Å^3$).

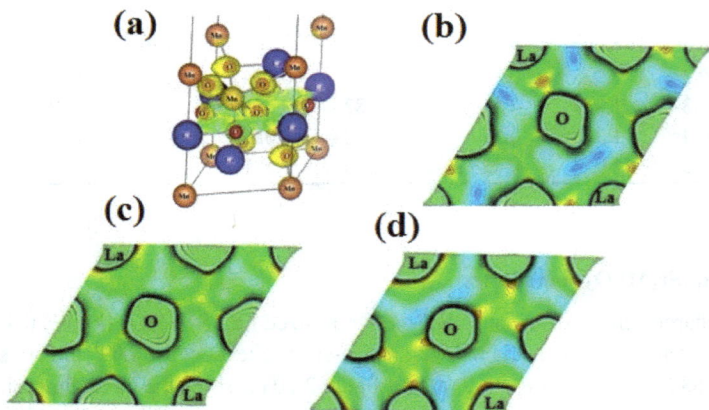

Figure 2.69 (a) 3D unit cell of $La_{1-x}Sr_xMnO_3$, with (004) plane shaded. Two dimensional electron density distribution on the (004) plane for $La_{1-x}Sr_xMnO_3$, **(b)** x=0.3, **(c)** x=0.4 and **(d)** x=0.5 (contour range: 0-1.0 $e/Å^3$, contour interval: 0.05 $e/Å^3$).

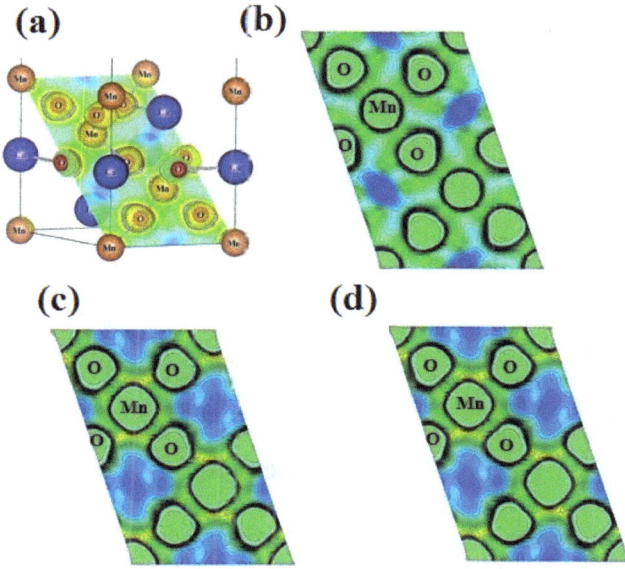

Figure 2.70 (a) *3D unit cell of La$_{1-x}$Sr$_x$MnO$_3$, with (012) plane shaded. Two dimensional electron density distribution on the (012) plane for La$_{1-x}$Sr$_x$MnO$_3$, **(b)** x=0.3, **(c)** x=0.4 and **(d)** x=0.5 (contour range: 0-1.0 e/Å^3, contour interval: 0.05 e/Å^3).*

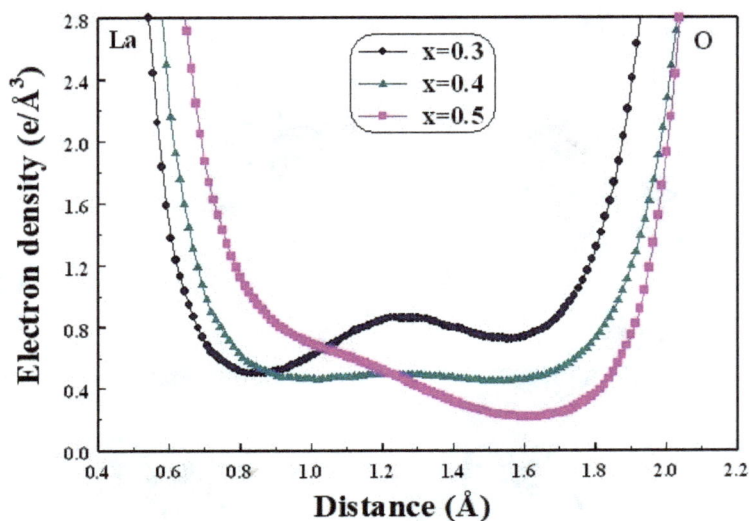

Figure 2.71 One dimensional electron density profiles along the La and O atoms in La$_{1-x}$Sr$_x$MnO$_3$, x=0.3, 0.4 and 0.5.

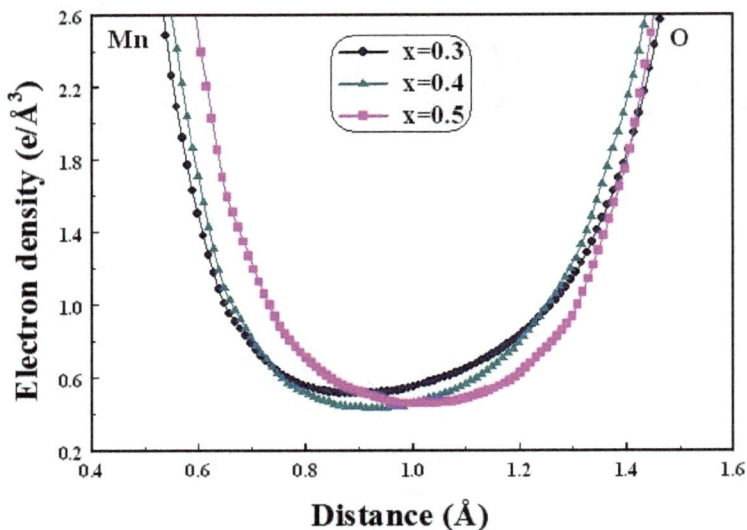

Figure 2.72 One dimensional electron density profiles along the Mn and O atoms in La$_{1-x}$Sr$_x$MnO$_3$, x=0.3, 0.4 and 0.5.

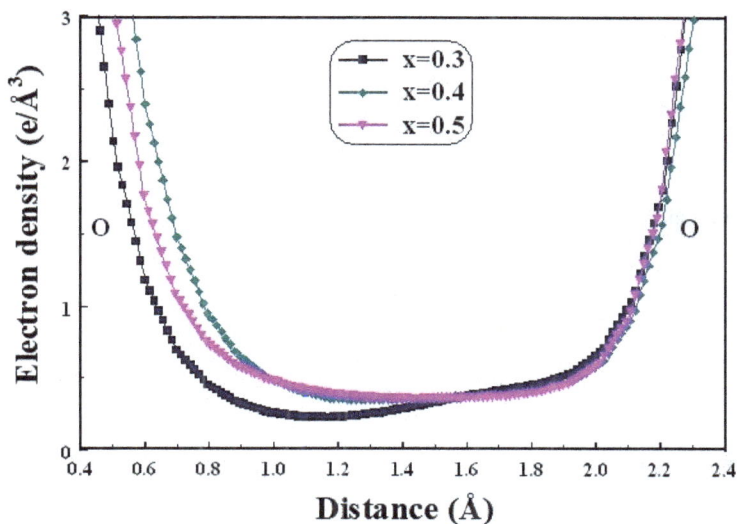

Figure 2.73 *One dimensional electron density profiles along the O and O atoms in La$_{1-x}$Sr$_x$MnO$_3$, x=0.3, 0.4 and 0.5.*

Table 2.29 *Bond lengths and mid bond electron densities of La-O, Mn-O and O-O bonds for La$_{1-x}$Sr$_x$MnO$_3$, x= 0.3, 0.4 and 0.5.*

Samples	Bonding					
	La-O		Mn-O		O-O	
	Bond length (Å)	Mid bond electron density (e/Å^3)	Bond length (Å)	Mid bond electron density (e/Å^3)	Bond length (Å)	Mid bond electron density (e/Å^3)
x=0.3	2.510	0.507	1.954	0.530	2.783	0.283
x=0.4	2.506	0.451	1.953	0.440	2.779	0.347
x=0.5	2.489	0.214	1.944	0.456	2.760	0.356

References

[1]   Collins D. M., Nature, 298, 49 (1982). https://doi.org/10.1038/298049a0

[2]   Momma K, Izumi F, VESTA: a three-dimensional visualization system for electronic and structural analysis. J. Appl. Crystallogr., 41, 653 (2008). https://doi.org/10.1107/S0021889808012016

[3]   Petricek V, Dusek M, Palatinus L, Kristallogr Z, Crystallographic Computing System JANA2006: General features, 229, 345 (2014)

[4]   Rietveld H.M., J. Appl. Crystallogr., 2, 65 (1969). https://doi.org/10.1107/S0021889869006558

[5]   Ruben A. D., Fujio I., Superfast program PRIMA for the Maximum Entropy Method, Advanced Materials Laboratory, National Institute for Material Science, Ibaraki, Japan (2004), 3050044

[6]   Wood D. L, Tauc J, Phys Rev B., 5, 3144 (1972). https://doi.org/10.1103/PhysRevB.5.3144

Chapter 3

Analysis of Results

Abstract

Chapter 3 deals with the analysis of the results obtained from the powder X-ray diffraction, scanning electron microscopy, energy dispersive X-ray spectroscopy, UV-visible spectroscopy and vibrating sample magnetometry for all the synthesized doped lanthanum chromite and manganite materials. The qualitative and quantitative MEM charge density distribution has been discussed for all the synthesized samples. An effort has also been made to correlate the structural and magnetic properties.

Keywords

Charge Density, Ferromagnetism, MEM, Energy Dispersive Analysis, Scanning Electron Microscopy, Optical Studies

Contents

3.1 Introduction

In this chapter, the results obtained from various characterization techniques such as powder X-ray diffraction (XRD), UV-visible absorption spectra (UV-vis), scanning electron microscopy (SEM), energy dispersive X-ray spectroscopy (EDS) and vibrating sample magnetometry (VSM), for all the synthesized doped lanthanum chromites and lanthanum manganites have been analyzed in a detailed manner. In addition, the charge density distribution obtained by the maximum entropy method (MEM) [Collins, 1982] for all the samples has been discussed in detail.

3.2 Synthesis of manganite structure materials

In the present work, materials with manganite structure such as doped lanthanum chromites and lanthanum manganites have been synthesized with the high temperature solid state reaction method. During the synthesis procedure, the various steps like, calcination, sintering and multiple grinding have been followed. Table 3.1 gives the calcination temperatures, sintering temperatures and grinding duration for the synthesis of lanthanum chromite and lanthanum manganite materials.

Table 3.1 *Calcination temperatures, sintering temperatures and grinding duration for the synthesis of doped lanthanum chromites and lanthanum manganites*

Samples	Calcination		Grinding duration (h)	I Sintering		Regrinding duration (h)	II Sintering	
	Temp. (°C)	Dura-tion (h)		Temp. (°C)	Dura-tion (h)		Temp. (°C)	Dura-tion (h)
LCCCMn	1000	4	-	1500	6	-	-	-
LCCCFe	1000	4	-	1500	6	-	-	-
LCCCCu	1000	4	-	1500	6	-	-	-
LCMO	1300	5	3	1400	12	3	1450	15
LSMO	1250	8	5	1400	12	3	1450	15

LCCCMn - $(La_{0.8}Ca_{0.2})(Cr_{0.9-x}Co_{0.1}Mn_x)O_3$
LCCCFe - $(La_{0.8}Ca_{0.2})(Cr_{0.9-x}Co_{0.1}Fe_x)O_3$
LCCCCu - $(La_{0.8}Ca_{0.2})(Cr_{0.9-x}Co_0 Cu_x)O_3$
LCMO - $La_{1-x}Ca_xMnO_3$
LSMO - $La_{1-x}Sr_xMnO_3$

133

3.3   Structural analysis using powder X-ray diffraction

The synthesized lanthanum chromite and lanthanum manganite samples have been characterized using powder X-ray diffraction for their structural properties. To investigate the structural properties in a detailed manner, powder profile refinement was carried out with the Rietveld method [Rietveld, 1969] using XRD data sets with the software program JANA 2006 [Petříček, 2006]. During the refinement, the cell constants, peak shift, preferred orientation, background profile shape and other parameters were refined and hence the error difference between the observed XRD profiles and the theoretically modeled profiles was minimized.

3.3.1   (Co, Mn) doped (La, Ca) based chromites - $(La_{0.8}Ca_{0.2})(Cr_{0.9-x}Co_{0.1}Mn_x)O_3$

The powder X-ray diffraction patterns for the synthesized chromite samples $(La_{0.8}Ca_{0.2})(Cr_{0.9-x}Co_{0.1}Mn_x)O_3$, (x=0.03, 0.06, 0.09 and 0.12) are shown in figure 2.1 (a). Figure 2.1 (b) shows the enlarged X-ray diffraction patterns for the (121), the (220) and the (040) Bragg planes. The powder X-ray diffraction patterns indicate a single phase of the prepared samples. All the peaks in XRD patterns match well with the JCPDS file number 33-0701. Shifting of Bragg peaks for various planes for all Mn concentrations (x=0.03, 0.06, 0.09 and 0.12) of the samples have been listed in table 2.1 as 2θ values. The enlarged XRD patterns (figure 2.1 (b)) and table 2.1 indicate that the Bragg peaks are shifted first (for x=0.06) towards the right, and then left (for x=0.09) for the increasing Mn incorporation in the Cr site of the lattice. This means that, for the prepared chromite samples, there is no systematic variation in the unit cell volume for different doping concentration of Mn at the lattice site of Cr.

For the synthesized chromite samples, the variation in the unit cell volume with different doping compositions can be explained by the transformation of the doped cations (Cr, Co and Mn) into their higher valence states of smaller ionic radii and vice-versa. In the co-doped lanthanum chromite systems, when Co^{3+} trivalent cations are doped, Co^{3+} may transform into high and low valence cations as $2Co^{3+} <=> Co^{2+} + Co^{4+}$ (ionic radius of Co^{2+} is 0.745 Å, that for Co^{3+} is 0.61 Å and that for Co^{4+} is 0.53 Å) [Shannon, 1976], and since this transformation is a reversible process, there is no systematic variation in the unit cell volume for the increasing Mn concentration [Gupta and Whang, 2007]. Also, the change of divalent cation of Ca^{2+} into trivalent cation of La^{3+}, leads to a charge compensating transition of Cr^{3+} to Cr^{4+} ions. Due to this transition, the Cr-O bond contracts and hence the unit cell volume decreases. Hence, in the synthesized co-doped system, the trivalent cations Co^{3+} and Cr^{3+} transform into low and high valence cations [Gupta and Whang, 2007] (ionic radius of Cr^{3+} is 0.62 Å and that for Cr^{4+} is 0.55 Å) [Shannon, 1976]. Since the high valence cations have low ionic radii and vice versa, for various doping concentration (x=0.03, 0.06, 0.09 and 0.12) of Mn, there is no systematic variation in the unit cell volume taking place.

For the prepared samples, the refinement was performed by considering the orthorhombic crystal structure model which has the space group of *Pnma* (Space group No: 62) with four molecules per unit cell. The atomic positional coordinates (x, y, z) were taken as (0.0267, 0.25, -0.004) for La/Ca, (0, 0, 0.5) for Cr/Co/Mn, (0.4905, 0.25, 0.0684) for apex oxygen O1 and (0.2193, 0.5361, 0.2165) for planar oxygen O2 [Brajesh et al., 2015, Ong et al., 2008]. The refined XRD profiles of $(La_{0.8}Ca_{0.2})(Cr_{0.9-x}Co_{0.1}Mn_x)O_3$, (x=0.03, 0.06, 0.09 and 0.12) are shown in figures 2.2 (a) - (d). In these figures, the observed XRD profiles are indicated by '××' symbols and the calculated XRD profiles are indicated by solid lines. At the bottom of the plots, the difference between the observed and calculated profiles is shown. The refined structural parameters and the reliability indices for the prepared samples are presented in table 2.2. The reliability indices indicate a better fitting between the observed and calculated profiles for all the synthesized samples.

3.3.2 (Co, Fe) doped (La, Ca) based chromites - $(La_{0.8}Ca_{0.2})(Cr_{0.9-x}Co_{0.1}Fe_x)O_3$

Figure 2.4 (a) shows the raw powder XRD profiles for the synthesized (Co, Fe) doped (La, Ca) based chromites $(La_{0.8}Ca_{0.2})(Cr_{0.9-x}Co_{0.1}Fe_x)O_3$, (x=0.03, 0.06, 0.09 and 0.12). The XRD patterns show sharp peaks, which confirm that the prepared samples have been well crystallized. The prominent peaks of the XRD patterns match well with the Joint Committee on Powder Diffraction Standards (JCPDS) database PDF No. 33-0701 and confirm the single phase perovskite structure of the samples. Figure 2.4 (b) shows the enlarged XRD peaks for the (121) plane for all the prepared samples and indicates that the intensity of the XRD peak corresponding to the Fe concentration of x=0.03 is smaller than that of all other compositions (x=0.06, 0.09 and 0.12). There is no considerable change in intensity for the Fe concentration of x=0.06, 0.09 and 0.12. From figure 2.4 (b), it is observed that, the peaks corresponding to the Fe concentration of x=0.06 and 0.09 shifts towards the lower angle side with respect to the x=0.03 concentration and this left shift can be explained in terms of Fe^{2+} ions (0.78 Å) having larger ionic radii replacing the Cr^{3+} (0.615 Å) ions [Shannon, 1976]. But for x=0.12 sample, the peak shifts towards the higher angle side with respect to the x=0.03 concentration and the reason for this right shift can be attributed to the existence of Fe^{3+} and Cr^{4+}, arising due to charge compensating transition [Gupta and Whang, 2007].

The crystal structure refinement of the synthesized doped chromite samples $(La_{0.8}Ca_{0.2})(Cr_{0.9-x}Co_{0.1}Fe_x)O_3$ (x=0.03, 0.06, 0.09 and 0.12) was performed in the orthorhombic setting with a space group of *Pnma* (Space group No: 62) with four formula unit per unit cell. In the orthorhombic crystal structure setting, the atomic coordinates were fixed at

(0.0267, 0.25, -0.004) for La/Ca atoms, (0,0,0.5) for Cr/Co/Fe atoms, (0.4905, 0.25, 0.0684) for apex O1 atoms and (0.2193, 0.5361, 0.2165) for planar O2 atoms [Brajesh et al., 2015, Ong et al., 2008]. The refined XRD powder profiles for the synthesized (Co, Fe) doped (La, Ca) based chromites are shown in figures 2.6 (a) - (d). Results obtained from the Rietveld refinement [Rietveld, 1969] are presented in table 2.3. The reliability factors and great fit in table 2.3 indicate a good fit between the observed and calculated XRD patterns.

3.3.3 (Co, Cu) doped (La, Ca) based chromites - $(La_{0.8}Ca_{0.2})(Cr_{0.9-x}Co_{0.1}Cu_x)O_3$

The powder XRD patterns of (Co, Cu) doped (La, Ca) based chromite samples $(La_{0.8}Ca_{0.2})(Cr_{0.9-x}Co_{0.1}Cu_x)O_3$, (x=0.00, 0.03 and 0.12) are shown in figure 2.7 (a) and the peaks for the planes (121) and (040) are enlarged for comparison and are presented in figures 2.7 (b) and 2.7 (c) respectively. The powder XRD patterns confirm that the synthesized samples are well crystallized and there are no additional phases present in the samples. The 2θ positions of all XRD peaks were indexed to $LaCrO_3$ with Joint Committee on Powder Diffraction Standards (JCPDS) PDF #33-0701. The synthesized samples crystallize into orthorhombic structure with space group of *Pnma* (Space group No: 62). Figures 2.7 (b) and 2.7 (c) show that the Bragg peaks shift towards the higher angle side from the undoped (x=0.00) sample for the increasing concentration of copper. The reason for this right shift can be explained in terms of Cu^{3+} ions (0.54 Å) having smaller ionic radii replacing the Cr^{3+} (0.615 Å) [Shannon, 1976] ions. There may be another reason for this right shift. Substitution of divalent cations at the La site or Cr site or both in lanthanum chromite, results in a charge compensating transition of trivalent chromium ions (Cr^{3+}) into tetravalent chromium ions (Cr^{4+}) ions. This transformation increases the concentration of Cr^{4+}, which leads to the contraction of Cr-O bonds and decreases the volume of the unit cell [Gupta and Whang, 2007].

For the synthesized chromite samples, the structural refinement was done by considering the orthorhombic crystal structure of the prepared samples with four molecules per unit cell with space group *Pnma* (Space group No: 62). For the orthorhombic structure, the atomic positional coordinates for La/Ca atom were fixed at (0.0267,0.25, -0.004), Cr/Co/Cu atoms are fixed at (0, 0, 0.5), O1 apex atoms are fixed at (0.4905, 0.25, 0.0684) and O2 planar atoms are fixed at (0.2193, 0.5361, 0.2165) [Brajesh et al., 2015, Ong et al., 2008]. The fitted profiles for (Co, Cu) doped (La, Ca) based chromites are shown in the figures 2.9 (a) - (c). The refined structural parameters are listed in table 2.4 and it clearly shows that, incorporation of Cu at the Cr site of the lattice decreases the lattice parameters and cell volume. This is due to the Cu^{3+} ions having smaller ionic radius (0.54

Å) replacing the Cr^{3+} (0.615 Å) ions. The reliability indices and quality of fit given in table 2.4, show the perfectness of the refinement.

The prepared (Co, Mn), (Co, Fe) and (Co, Cu) doped (La, Ca) based chromite samples have a orthorhombic crystal structure. The orthorhombic unit cells of the synthesized chromite samples have been obtained by the visualization software, Visualization for Electronic and STructural Analysis (VESTA) [Momma and Izumi, 2008] which are shown in figures 2.3, 2.5 and 2.8. They consist of twenty atoms (4 La atoms, 4 Cr atoms and 12 O atoms) per unit cell. These orthorhombic unit cells have corner linked octahedra CrO_6 in which Cr atoms occupy the center position and O atoms occupy the vertex position of the octahedra. La atoms occupy the space between the CrO_6 octahedra.

3.3.4 $La_{1-x}Ca_xMnO_3$ manganites

Figure 2.10 (a) shows the powder X-ray diffraction patterns for $La_{1-x}Ca_xMnO_3$ (x=0.1, 0.2, 0.3, 0.4 and 0.5), which indicate the single phase of the samples. The XRD peaks for all the samples match well with the JCPDS PDF # 72-0842 and confirm the orthorhombic structure of the samples with *Pnma* space group (Space Group No:62). The intensity of the XRD peaks decreases for the increasing incorporation of Ca at La site of the lattice. The reason for this decreasing intensity may due to the fact that the atomic number of Ca (Z=20) which replaces La, is lesser than the atomic number of lanthanum (Z=57).

Figure 2.10 (b) shows the enlarged XRD patterns for the synthesized samples corresponding to the (121) and (040) planes, which indicates the right shift of the diffraction peaks for all the Ca doped lanthanum manganite samples. The reason for this right shift is due to the presence of Mn^{4+} ions in $LaMnO_3$ by the substitution of Ca at the La site of the lattice. Actually, the two cations La^{3+} and Ca^{2+} have very slight difference in their ionic radii (1.36 Å and 1.34 Å) [Shannon, 1976] and hence one should expect negligible shifts in X-ray diffraction peaks. But figure 2.10 (b) shows a clear right shift and the reason for this right shift can be explained as follows. Substitution of Ca^{2+} at La^{3+} site converts the valency of Mn (as Mn^{3+} and Mn^{4+}) to maintain the charge neutrality of the compound [Tyson et al., 1999]. Table 3.2 displays the valence of Mn for the increasing doping concentration of Ca at the La site of $LaMnO_3$. As the Ca^{2+} content increases, Mn^{4+} content also increases. Hence, by the substitution of Ca^{2+} at La^{3+}, the effective ionic radii of La^{3+} and Mn^{3+} decreases and therefore, the right shift has been observed in the XRD patterns. The disappearance of the (101) peak takes place for x=0.5, which was also observed by Lira-Hernantez et al., [2010].

Table 3.2 *Valence of manganese (Mn) ion with respect to calcium (Ca) doping in lanthanum manganite ($La_{1-x}Ca_xMnO_3$).*

Calcium doping concentration (x)	Valence of Mn	Valence	
		Mn^{3+} (%)	Mn^{4+} (%)
0.0	3+	100	0
0.1	3.1+	90	10
0.2	3.2+	80	20
0.3	3.3+	70	30
0.4	3.4+	60	40
0.5	3.5+	50	50
0.6	3.6+	40	60
0.7	3.7+	30	70
0.8	3.8+	20	80
0.9	3.9+	10	90
1.0	4+	0	100

The refinement was done for the prepared samples by considering the orthorhombic structure of space group *Pnma* with four molecules per unit cell. Figure 2.11 shows the distorted orthorhombic unit cell structure for the $La_{0.8}Ca_{0.2}MnO_3$ sample, which was obtained through the visualization software VESTA [Momma and Izumi, 2008]. This unit cell has corner linked MnO_6 octahedra which consists of oxygen atoms at their vertices and Mn atoms at their centers. Lanthanum atoms occupy the space between the MnO_6 octahedra. The fitted profiles for $La_{1-x}Ca_xMnO_3$ (x=0.1, 0.2, 0.3, 0.4 and 0.5) samples are shown in figures 2.12 (a) - (e). Table 2.5 gives the refined structural parameters for the grown samples, which indicates the shrinkage in the unit cell for the increasing incorporation of calcium. This shrinkage is due to the enhancement of Mn^{4+} ions in lanthanum manganite by the doping of calcium. Reliability indices and good fit in table 2.5 indicate that there is a perfect fit between the observed and the calculated XRD profiles.

3.3.5 $La_{1-x}Sr_xMnO_3$ manganites

The raw XRD patterns of $La_{1-x}Sr_xMnO_3$, (x=0.3, 0.4 and 0.5) manganites are shown in figure 2.13 (a), which confirms the single phase perovskite structure of the prepared samples. The XRD data for all samples match well with the JCPDS PDF # 47-0444 and confirm the rhombohedral structure of the synthesized samples. Figure 2.13 (b) shows the enlarged XRD peaks for the planes (110) and (202) and it is observed that all the Bragg peaks shift towards the higher angle side for the increasing incorporation of Sr. The ionic radii of La^{3+} and Sr^{2+} are 1.36 Å and 1.44Å [Shannon, 1976] and hence one should expect a left shift in the XRD peaks. But, the observed right shift was due to the existence of Mn^{4+} in $LaMnO_3$, by the substitution of Sr. When Sr^{2+} is doped at the La^{3+} site of $LaMnO_3$, the Mn ions have their valences as Mn^{3+} and Mn^{4+} to attain the charge neutrality in the system [Szytuła, 2010]. Hence, for the increasing incorporation of Sr^{2+} in the La^{3+} site of $LaMnO_3$, the amount Mn^{4+} also increases, which contracts the Mn-O bond. Also, Elemans et al. [1971], reported that, divalent cations substitution on lanthanum contract the Mn-O bond length. Hence, shrinkage in the unit cell is observed when strontium ions are substituted in the lanthanum site of lanthanum manganite.

For the Sr doped $LaMnO_3$ samples, the refinement was carried out with a rhombohedral structure model having space group $R\bar{3}c$ (space group No: 167) with six molecules per unit cell [Israel et al., 2012]. In the rhombohedral structure with hexagonal setting, the atomic position coordinates for the La and Sr atoms were assumed as (0, 0, 0.25), for the Mn atoms as (0, 0, 0) and for the O atoms as (0.4561, 0, 0.25). In the rhombohedral structure, the MnO_6 octahedron consists of a Mn atom at its center and O atoms at its vertices. The MnO_6 octahedron share its vertices with neighboring octahedron to form the three dimensional network. The refined profiles of the manganite samples $La_{1-x}Sr_xMnO_3$, (x=0.3, 0.4 and 0.5) are shown in figures 2.14 (a) - (c) and the refined structural parameter are given in table 2.6. The cell parameters 'a' and 'b' decrease linearly for the increasing concentration of Sr. But, the cell parameter 'c' does not show any systematic variations with Sr composition. Hammouche et al. observed the same behavior for the Sr doped $LaMnO_3$ samples [Hammouche et al. 1989]. The reliability indices shown in table 2.6 evidences that, successful refinement was carried out for the prepared samples.

The refined structural parameters for all the doped lanthanum chromites and lanthanum manganites are given in table 3.3. For $(La_{0.8}Ca_{0.2})(Cr_{0.9-x}Co_{0.1}Mn_x)O_3$ and $(La_{0.8}Ca_{0.2})(Cr_{0.9-x}Co_{0.1}Fe_x)O_3$ chromites, there is no systematic variations in lattice constants and unit cell volume with the doping concentration of Mn and Fe. For (Co, Cu) doped (La, Ca) based chromites $(La_{0.8}Ca_{0.2})(Cr_{0.9-x}Co_{0.1}Cu_x)O_3$ with x=0.00, 0.03 and 0.12, the lattice constants and the unit cell volume decrease by the incorporation of Cu at the lattice site of Cr. For Ca and Sr doped lanthanum manganites $(La_{1-x}(Ca_x/Sr_x)MnO_3)$, the lattice constants and the unit cell volume decrease for the increasing incorporation of Ca/Sr at the lattice site of La.

Table 3.3 *Refined structural parameters for doped lanthanum chromites and lanthanum manganites.*

Samples	Crystallo-graphic system	Conc. (x)	a (Å)	b (Å)	c (Å)	Unit cell volume(Å^3)
LCCCMn	Ortho-rhombic (*Pnma*)	0.03	5.474(7)	7.770(11)	5.524(8)	235.00(8)
		0.06	5.469(9)	7.767(12)	5.522(11)	234.60(18)
		0.09	5.512(8)	7.799(8)	5.504(8)	236.65(6)
		0.12	5.474(4)	7.753(6)	5.510(4)	233.91(5)
LCCCFe	Ortho-rhombic (*Pnma*)	0.03	5.516(3)	7.772(7)	5.481(3)	235.05(3)
		0.06	5.507(6)	7.784(5)	5.507(4)	236.13(6)
		0.09	5.471(5)	7.747(6)	5.506(6)	233.43(5)
		0.12	5.514(6)	7.757(8)	5.478(6)	234.34(7)
LCCCCu	Ortho-rhombic (*Pnma*)	0.00	5.517(3)	7.792(8)	5.551(2)	238.68(8)
		0.03	5.484(9)	7.769(7)	5.511(3)	234.82(5)
		0.12	5.459(9)	7.724(5)	5.466(2)	230.23(4)
LCMO	Ortho-rhombic (*Pnma*)	0.1	5.508(5)	7.747(5)	5.484(5)	234.07(6)
		0.2	5.482(9)	7.737(9)	5.483(9)	233.23(6)
		0.3	5.456(3)	7.710(3)	5.472(3)	230.24(4)
		0.4	5.431(12)	7.666(12)	5.444(12)	226.69(7)
		0.5	5.423(8)	7.632(8)	5.415(8)	224.18(5)
LSMO	Rhombo-hedral (*R$\bar{3}$c*)	0.3	5.504(5)	5.504(5)	13.355(4)	350.37(3)
		0.4	5.495(11)	5.495(11)	13.378(3)	349.81(15)
		0.5	5.457(6)	5.457(6)	13.365(5)	344.83(4)

Conc. - Concentration

LCCCMn - $(La_{0.8}Ca_{0.2})(Cr_{0.9-x}Co_{0.1}Mn_x)O_3$

LCCCFe - $(La_{0.8}Ca_{0.2})(Cr_{0.9-x}Co_{0.1}Fe_x)O_3$

LCCCCu - $(La_{0.8}Ca_{0.2})(Cr_{0.9-x}Co_{0.1}Cu_x)O_3$

LCMO - $La_{1-x}Ca_xMnO_3$

LSMO - $La_{1-x}Sr_xMnO_3$

3.4 Grain size analysis

The average grain sizes of the prepared lanthanum chromite and lanthanum manganite samples have been determined from the observed full width at half maximum of the XRD peaks using the Scherrer formula [Culllity, 2001], $t = 0.9\lambda/(\beta \cos\theta)$, where t is the grain size (size of the coherently diffracting domain), λ is wavelength of X-ray used, β is the full width at half maximum and θ is the Bragg angle, through the GRAIN software [Saravanan, 2008].

The average grain size for the (Co, Mn) doped (La, Ca) based chromites $(La_{0.8}Ca_{0.2})(Cr_{0.9-x}Co_{0.1}Mn_x)O_3$, (x=0.03, 0.06, 0.09, 0.12) ranges from 19 nm and 31 nm.

The average grain size for the (Co, Fe) doped (La, Ca) based chromites, $(La_{0.8}Ca_{0.2})(Cr_{0.9-x}Co_{0.1}Fe_x)O_3$, (x=0.03, 0.06, 0.09 and 0.12) ranges from 17 nm to 21 nm.

The average grain size for the (Co, Cu) doped (La, Ca) based chromites, $(La_{0.8}Ca_{0.2})(Cr_{0.9-x}Co_{0.1}Cu_x)O_3$, (x=0.00, 0.03 and 0.12) ranges from 13 nm to 17 nm.

The average grain sizes for the synthesized $La_{1-x}Ca_xMnO_3$ (x=0.1, 0.2, 0.3, 0.4 and 0.5) samples are found to be in the range of 35 nm to 63 nm.

The average grain size for the synthesized $La_{1-x}Sr_xMnO_3$ (x=0.3, 0.4 and 0.5) samples are found to be in the range of 10 nm to 15 nm.

The average grain size range estimated from Scherrer formula [Culllity, 2001] for all the doped lanthanum chromites and lanthanum manganites have been presented in table 3.4.

The final sintered pellet samples have been forcefully ground to get fine powder samples for XRD characterization. Due to this reason, the grain size estimated for synthesized powder samples range in the order of nanometers.

Table 3.4 *Average grain size range of doped lanthanum chromites and lanthanum manganites from XRD*

Samples	Average grain size range (nm)
LCCCMn	19 - 31
LCCCFe	17 - 21
LCCCCu	13 - 17
LCMO	35 - 63
LSMO	10 - 15

LCCCMn - $(La_{0.8}Ca_{0.2})(Cr_{0.9-x}Co_{0.1}Mn_x)O_3$
LCCCFe - $(La_{0.8}Ca_{0.2})(Cr_{0.9-x}Co_{0.1}Fe_x)O_3$
LCCCCu - $(La_{0.8}Ca_{0.2})(Cr_{0.9-x}Co_{0.1}Cu_x)O_3$
LCMO - $La_{1-x}Ca_xMnO_3$
LSMO - $La_{1-x}Sr_xMnO_3$

3.5 Surface morphology and elemental analysis by SEM/EDS

The surface morphology of the synthesized lanthanum chromite and lanthanum manganite samples has been analyzed using the SEM images. For the lanthanum chromite samples, the SEM images were recorded for different magnifications ($\times 1500$, $\times 5000$, $\times 10,000$).

3.5.1 (Co, Mn) doped (La, Ca) based chromites - $(La_{0.8}Ca_{0.2})(Cr_{0.9-x}Co_{0.1}Mn_x)O_3$

Figures 2.15 (a) - (d) show the SEM images of (Co, Mn) doped (La, Ca) based chromite samples for a magnification of $\times 10,000$. The SEM images reveal that the particles are spherical in shape and have been distributed without much agglomeration. The SEM particle size ranges from 0.35 μm to 0.66 μm.

3.5.2 (Co, Fe) doped (La, Ca) based chromites - $(La_{0.8}Ca_{0.2})(Cr_{0.9-x}Co_{0.1}Fe_x)O_3$

Figures 2.16 (a) - (d) show the surface morphological SEM images with magnification $\times 10,000$ for (Co, Fe) doped (La, Ca) based chromite samples. Particles of spherical shape with different sizes have been heterogeneously distributed in all the samples. The average particle size for the synthesized samples ranges from 0.36 μm to 0.58 μm.

The EDS spectra of $(La_{0.8}Ca_{0.2})(Cr_{0.9-x}Co_{0.1}Fe_x)O_3$ (x=0.03, 0.06, 0.09 and 0.12) samples are presented in figures 2.17 (a) - (d). The various peaks in the spectra indicate the different atomic species (La, Ca, Cr, Co, Fe and O) present in the samples. The extra carbon peaks are also noted in the EDS spectra for all samples which are due to the carbon paste used for mounting the samples. The atomic percentage and weight percentage of the various elements present in the synthesized samples are given in table 2.7 which confirms that no other impurities are present in the prepared samples.

3.5.3 (Co, Cu) doped (La, Ca) based chromites - $(La_{0.8}Ca_{0.2})(Cr_{0.9-x}Co_{0.1}Cu_x)O_3$

SEM images of the (Co, Cu) doped (La, Ca) based chromites with a magnification of $\times 10,000$ are displayed in the figures 2.18 (a) - (c). The surface morphological images of samples with x=0.00 and 0.03 show the presence of tiny individual particles of different sizes. These particles have been distributed uniformly without much agglomeration. But, the surface morphological image of the sample with x=0.12 shows that the particles of different sizes have been distributed heterogeneously with voids. The average particle size of prepared (Co, Cu) doped (La, Ca) based chromites range from 0.23 μm to 0.28 μm.

The EDS spectra of $(La_{0.8}Ca_{0.2})(Cr_{0.9-x}Co_{0.1}Cu_x)O_3$ (x=0.00, 0.03, and 0.12) samples are presented in figures 2.19 (a) - (c). The peaks corresponding to La, Ca, Cr, Co, Cu and O atoms are shown in the EDS spectra, which indicate the purity of the sample. The additional carbon peak is also observed for all the prepared samples which may be due to the carbon paste used for sample mounting. The atomic and weight percentages of the elemental compositions are given in table 2.8 and it again confirms that, there are no other impurities present in the samples.

3.5.4 $La_{1-x}Ca_xMnO_3$ manganites

Figures 2.20 (a) - (e) show SEM images of the synthesized Ca doped lanthanum manganite $La_{1-x}Ca_xMnO_3$ (x=0.1, 0.2, 0.3, 0.4 and 0.5) samples with a magnification of ×10000. The SEM pictures for all the prepared samples clearly show that the particles are in polygonal form with different sizes [Ewe et al., 2012]. These particles are highly aggregated with no voids. The size of the polygonal particles increases gradually for Ca doping concentration up to x=0.3. Then, for x=0.4 sample, a slight decrease in size of the particle is observed. The smaller aggregated particles with no voids are seen in the SEM image of the sample with x=0.5. Hence, there is a significant change in the particle size for the different doping concentration of calcium.

The EDS spectra for the Ca doped lanthanum manganite $La_{1-x}Ca_xMnO_3$ (x=0.1, 0.2, 0.3, 0.4 and 0.5) samples are shown in figures 2.21 (a) - (e). The EDS spectra show the various characteristic peaks corresponding to the La, the Ca, the Mn and the O atoms. Table 2.9 gives the atomic and weight percentages of the La, Ca, Mn and O atoms present in the sample. The atomic percentages of the prepared Ca doped $LaMnO_3$ samples match closely with the stoichiometry of the samples and it confirms the purity of the samples.

3.5.5 $La_{1-x}Sr_xMnO_3$ manganites

The SEM micrographs of strontium doped lanthanum manganite samples recorded with a magnification of ×10000 are shown in figures 2.22 (a) - (c). Aggregated particles in polygonal shape with no voids are observed from the SEM images of all the Sr doped samples. The particles have clear grain boundary between them. The particle size gradually decreases for the increasing incorporation of Sr at the La site of $LaMnO_3$. Hence, the SEM images reveal a significant change in particle size for the different composition of Sr.

Figures 2.23 (a) - (c) show the EDS spectra of Sr doped lanthanum manganite samples and the peaks corresponding to La, Sr, Mn and O atoms are observed in each spectrum. The atomic and weight percentages of La, Sr, Mn and O atoms present in the synthesized samples are given in table 2.10 and confirms that, no other impurities are present in the samples.

The morphology and average particle size from the SEM images, for all the doped lanthanum chromites and lanthanum manganites have been tabulated in table 3.5. For the polycrystalline samples, the particle size from SEM may not necessarily be compared directly with grain size estimated from XRD. The grain size determined from powder XRD gives only the size of the coherently diffracting domains [Saravanan et al., 2010].

The average particle size of the lanthanum chromite samples ranges from 0.23 μm to 0.66 μm and the particles appear in spherical shape whereas the average particle size of the lanthanum manganite samples ranges from 2 μm to 9 μm and the particles appear in polygonal shape of different sizes.

Table 3.5 *The morphology and average particle size of doped lanthanum chromites and lanthanum manganites*

Samples	Morphology	Average particle size (μm)
LCCCMn	spherical	0.35 - 0.66
LCCCFe	spherical	0.36 - 0.58
LCCCCu	spherical	0.23 - 0.28
LCMO	polygonal	4-9
LSMO	polygonal	2-7

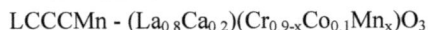

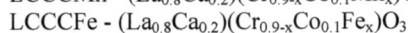

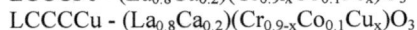

LCCCMn - $(La_{0.8}Ca_{0.2})(Cr_{0.9-x}Co_{0.1}Mn_x)O_3$
LCCCFe - $(La_{0.8}Ca_{0.2})(Cr_{0.9-x}Co_{0.1}Fe_x)O_3$
LCCCCu - $(La_{0.8}Ca_{0.2})(Cr_{0.9-x}Co_{0.1}Cu_x)O_3$
LCMO - $La_{1-x}Ca_xMnO_3$
LSMO - $La_{1-x}Sr_xMnO_3$

3.6 Optical band gap analysis by UV-visible absorption spectra

The optical band gap of synthesized lanthanum chromite and lanthanum manganite samples was evaluated from the UV-visible absorption spectra. The UV-visible absorption spectra were recorded in the range between 200 nm and 2000 nm. The energy band gap of the synthesized samples was estimated with the equation $\alpha h\nu = A (h\nu - E_g)^n$, where $h\nu$ is photon energy, α is absorbance and E_g is the energy band gap. This equation was proposed by Wood and Tauc [Wood and Tauc, 1972]. Using the Tauc equation [Wood and Tauc, 1972], a graph is plotted with $h\nu$ along the x-axis and $(\alpha h\nu)^2$ along the y-axis. The linear portion in Tauc plot [Wood and Tauc, 1972] extrapolated to $(\alpha h\nu)^2 = 0$ estimates the optical band gap.

3.6.1 (Co, Mn) doped (La, Ca) based chromites - $(La_{0.8}Ca_{0.2})(Cr_{0.9-x}Co_{0.1}Mn_x)O_3$

The optical band gap of $(La_{0.8}Ca_{0.2})(Cr_{0.9-x}Co_{0.1}Mn_x)O_3$, (x=0.03, 0.06, 0.09 and 0.12) samples was evaluated from the UV-visible absorption spectra shown in figure 2.24. The synthesized (Co, Mn) doped (La, Ca) based chromites are direct band gap materials for which n=1/2. The Tauc plot [Wood and Tauc, 1972], is shown in figure 2.25. For the prepared lanthanum chromite samples, the energy band gap values are given in table 2.11.

3.6.2 (Co, Fe) doped (La, Ca) based chromites - $(La_{0.8}Ca_{0.2})(Cr_{0.9-x}Co_{0.1}Fe_x)O_3$

The UV-vis absorption spectra of the synthesized $(La_{0.8}Ca_{0.2})(Cr_{0.9-x}Co_{0.1}Fe_x)O_3$, (x=0.03, 0.06, 0.09 and 0.12) samples are shown in figure 2.26. Figure 2.26 shows that the UV-visible absorption edges are at around 260 nm to 320 nm and the absorption peaks shift towards the higher wavelength side, which means the red shift, for the increasing incorporation of Fe at the lattice site of Cr. The optical band gap was evaluated from the Tauc plot [Wood and Tauc, 1972], which is shown in figure 2.27. The direct band gap values range from 2.135 eV to 2.405 eV and are presented in table 2.12 for various concentration of Fe. The optical band gap values are found to be decreasing for the increasing doping concentration of Fe at the lattice site of Cr.

3.6.3 (Co, Cu) doped (La, Ca) based chromites - $(La_{0.8}Ca_{0.2})(Cr_{0.9-x}Co_{0.1}Cu_x)O_3$

Figure 2.28 shows the UV-visible absorption spectra of the prepared (Co, Cu) doped (La, Ca) based chromites. Tauc plot [Wood and Tauc, 1972] is drawn for synthesized materials and shown in figure 2.29. The E_g values range from 1.859 eV to 2.448 eV and are given in table 2.13. With increasing incorporation of Cu at the Cr lattice site, the band gap energy is found to be decreasing. This behavior of decrease in the optical band gap has been attributed to the band gap narrowing due to s-d and p-d interactions and has been explained by the second order perturbation theory [Naseem et al., 2014].

3.6.4 $La_{1-x}Ca_xMnO_3$ manganites

Figure 2.30 shows the UV-visible absorption spectra of the synthesized $La_{1-x}Ca_xMnO_3$ (x=0.1, 0.2, 0.3, 0.4 and 0.5) manganite samples. Using the absorption data, a graph between (hv) and $(\alpha hv)^2$ is drawn as shown in figure 2.31 and the extrapolation of the linear portion of this graph to x-axis gives the band gap values. The optical band gaps for $La_{1-x}Ca_xMnO_3$ (x=0.1, 0.2, 0.3, 0.4 and 0.5) samples are given in table 2.14 and are found to be decreasing with the increasing concentration of Ca. This decrease in band gap is due to the small Jahn-Teller (JT) polaron distortion by Mn^{3+} ions [Jung et al., 1998;

Chaudhary et al., 2015]. The E_g values are almost closer to the reported values [Sultan and Ikram, 2015].

3.6.5 $La_{1-x}Sr_xMnO_3$ manganites

The UV-visible absorption spectra of $La_{1-x}Sr_xMnO_{3,}$ (x=0.3, 0.4 and 0.5) manganite samples are shown in figure 2.32. The absorption peaks of the synthesized samples are at around 260 nm to 270 nm. For the $La_{1-x}Sr_xMnO_3$ samples, the optical band gap has been determined by considering the indirect-allowed electronic transitions for which n=2. Figure 2.33 shows the Tauc plot [Wood and Tauc, 1972], which estimates the energy band gap of the prepared samples. The E_g values of the prepared Sr doped lanthanum manganites range from 2.442 eV to 2.487 eV and are tabulated in table 2.15. The variation in optical band gap values is attributed to the change in the ratio of Mn^{4+}/Mn^{3+} with respect to the ratio of La/Sr.

The optical band gap values estimated from the UV-visible absorption spectra, for all the doped lanthanum chromite and lanthanum manganite samples are given in table 3.6.

Table 3.6 Optical band gaps of doped lanthanum chromites and lanthanum manganites using UV-visible absorption spectra.

Samples	Optical band gap (eV)
LCCCMn	2.270 - 2.464
LCCCFe	2.135 - 2.405
LCCCCu	1.859 - 2.448
LCMO	1.411 - 1.739
LSMO	2.442 - 2.487

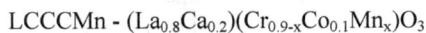

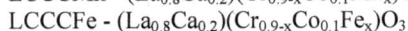

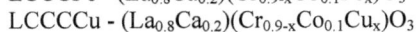

LCCCMn - $(La_{0.8}Ca_{0.2})(Cr_{0.9-x}Co_{0.1}Mn_x)O_3$
LCCCFe - $(La_{0.8}Ca_{0.2})(Cr_{0.9-x}Co_{0.1}Fe_x)O_3$
LCCCCu - $(La_{0.8}Ca_{0.2})(Cr_{0.9-x}Co_{0.1}Cu_x)O_3$
LCMO - $La_{1-x}Ca_xMnO_3$
LSMO - $La_{1-x}Sr_xMnO_3$

3.7 Magnetic analysis by vibrating sample magnetometer measurements

The magnetic properties of the synthesized lanthanum chromite and lanthanum manganite samples have been analyzed with the data recorded using a vibrating sample magnetometer. Room temperature M-H curves have been drawn for all the synthesized samples to study the magnetic behavior of the materials.

3.7.1 (Co, Mn) doped (La, Ca) based chromites - $(La_{0.8}Ca_{0.2})(Cr_{0.9-x}Co_{0.1}Mn_x)O_3$

Lanthanum chromite $(LaCrO_3)$ is a G-type antiferromagnetic material below 290 K and at 300 K, it is a poor electrical conducting material [Gonjal et al., 2013]. Reports on calcium doped lanthanum chromite show that, the Neel temperature T_N falls from 290 K to 160 K for the increasing calcium doping concentration and it exhibits weak ferromagnetism with large coercive fields. The weak antiferromagnetic behavior can be explained by canting of antiferromagnetically ordered Cr moments [Alvarez et al., 2008]. The room temperature M-H curves for the co-doped lanthanum chromites $(La_{0.8}Ca_{0.2})(Cr_{0.9-x}Co_{0.1}Mn_x)O_3$, (x=0.03, 0.06, 0.09 and 0.12) are shown in figure 2.34. The magnetic parameters such as saturation magnetization (M_s), remnant magnetization (M_r) and coercive field (H_c) are given in table 2.16. The non-saturation behavior of M-H curves with small value of saturation magnetization confirm that the synthesized samples exhibit predominant antiferromagnetic ordering due to Cr^{3+} spins [Terashita et al., 2012; Shukla et al., 2009]. The variation in saturation magnetization for various compositions (x=0.03, 0.06, 0.09 and 0.12) may be attributed to the magneto crystalline anisotropy of the samples.

3.7.2 (Co, Fe) doped (La, Ca) based chromites - $(La_{0.8}Ca_{0.2})(Cr_{0.9-x}Co_{0.1}Fe_x)O_3$

The M-H loops recorded for the $(La_{0.8}Ca_{0.2})(Cr_{0.9-x}Co_{0.1}Fe_x)O_3$, (x=0.03, 0.06, 0.09 and 0.12) samples at 300 K are shown in figure 2.35. The magnetic parameters are listed in table 2.17. The samples with x=0.03, 0.06 and 0.09, exhibit minor hysteresis loops with low saturation magnetization, which may be due to the magneto crystalline anisotropy and Cr^{3+} spin canting [Shukla et al., 2009]. But, the sample with x=0.12 exhibits large hysteresis with high saturated magnetization. This ferromagnetic behavior of x=0.12 sample can be explained in terms of disorder in antiferromagnetic interaction due to considerable increase of Fe concentration and associated uncompensated canted FM. By the substitution of Fe at the Cr lattice site, the usual Cr^{3+}- O^{2-}- Cr^{3+} networks are disturbed and the uncompensated Cr^{3+} ions yield additional canted FM [Tribedi and Ravi, 2013, Choudhry et al., 2015].

3.7.3 (Co, Cu) doped (La, Ca) based chromites - $(La_{0.8}Ca_{0.2})(Cr_{0.9-x}Co_{0.1}Cu_x)O_3$

The room temperature M-H curves of the $(La_{0.8}Ca_{0.2})(Cr_{0.9-x}Co_{0.1}Cu_x)O_3$, (x=0.00, 0.03 and 0.12) samples are shown in figure 2.36. The magnetic parameters such as M_s, M_r and H_c are given in table 2.18. The M-H curves shown in figure 2.36, do not show any hysteresis at all. Hence, the non-saturation behavior of M-H curve with observed small value of magnetization authenticates that all the synthesized samples are attributed to predominant antiferromagnetic ordering of Cr^{3+} spins [Terashita et al., 2012].

3.7.4 La$_{1-x}$Ca$_x$MnO$_3$ manganites

La$_{1-x}$Ca$_x$MnO$_3$ is a mixed valence system, in which Mn exist in two valence states such as Mn^{3+} and Mn^{4+}. Out of these two Mn ions, Mn^{3+} ion (3d^4) has one e$_g$ electron which can act as a mobile charge carrier in the compound and Mn^{4+} ion (3d^3) has empty e$_g$ orbitals (no charge carrier) [Loa et al., 2001]. The interaction between Mn^{3+} and Mn^{4+} ions via O^{2-} ions decides the magnetic property of Ca doped LaMnO$_3$. Usually, La$_{1-x}$Ca$_x$MnO$_3$ exhibits three magnetic phases such as antiferromagnetic insulating, ferromagnetic conducting and paramagnetic insulating phases. These three magnetic phases of La$_{1-x}$Ca$_x$MnO$_3$ depend on the dopant (Ca) concentration and temperature. At low temperatures, for $0 \leq x \leq 0.1$, the Ca doped lanthanum manganite samples behave as A-type antiferromagnetic material. For $0.1 \leq x \leq 0.21$, the compound behaves as a ferromagnetic insulating material and for $0.21 \leq x \leq 0.4$, the compound behaves as ferromagnetic conducting material. But, above x=0.5, the compound exhibits antiferromagnetic insulating behavior. The Curie temperature T$_C$ for La$_{1-x}$Ca$_x$MnO$_3$ (x=0.1-0.5) ranges from 180 K to 270 K. Beyond T$_C$, La$_{1-x}$Ca$_x$MnO$_3$ ($0 \leq x \leq 1$) samples exhibit paramagnetic insulating behavior [Ramirez, 1997].

The synthesized La$_{1-x}$Ca$_x$MnO$_3$ (x=0.1, 0.2, 0.3, 0.4 and 0.5) samples have been analyzed for their magnetic properties at 20 K and 300 K using vibrating sample magnetometry. The M-H loops recorded at 20 K are shown in figure 2.37. The samples with x=0.1, 0.2, 0.3 and 0.4 show ferromagnetic behavior. The magnetic parameters at 20 K are given in table 2.19. The saturation magnetization (M$_s$) gradually increases for the increasing incorporation of Ca at the lattice site of La. The ferromagnetic behavior of Ca doped LaMnO$_3$ samples can be explained by the canting of the antiferromagnetically ordered spins by the MnO$_6$ octahedral distortion and by double exchange interaction which takes place between Mn^{3+} and Mn^{4+} ions [Sultan and Ikram, 2015]. But for the sample with x=0.5, the M$_s$ abruptly decreases and behaves as a mixture of ferromagnetic and antiferromagnetic phases since this sample has equal concentration of Mn^{3+} and Mn^{4+} [Sultan and Ikram, 2015, Coey et al., 1999].

Figure 2.38 shows the M-H loops recorded at 300 K for the synthesized Ca doped LaMnO$_3$ samples and the magnetic parameters are presented in table 2.20. All the samples are found to be paramagnetic as they are showing linear behavior with M and H. At room temperature, (T=300K> T_c), there is a domination of paramagnetic nature of La sublattices [Sultan and Ikram, 2015] and the Mn spin fluctuations are so high that an applied field cannot align the spins [Jin et al., 1994]. Moreover, above T$_C$, more phonons are generated, which trap the charge carriers locally and hence, the material behave as an insulator. The transport behavior in the insulating phase can be discussed by adiabatic polaron hopping (APH) model and variable-range hopping model (VRH) [Thiesse et al.,

2015]. Table 2.20 shows that the paramagnetic magnetization also increases for the increasing doping concentration of Ca. But, for the sample with x=0.5, the saturation magnetization decreases which may be due to the existence of equal concentration of Mn^{3+} and Mn^{4+} ions.

3.7.5 $La_{1-x}Sr_xMnO_3$ manganites

$LaMnO_3$ is an A-type antiferromagnetic insulating material. But, substitution of divalent ions (like Sr^{2+} and Ca^{2+}) at La site changes its magnetic property. In Sr doped lanthanum manganites, Mn ions exist with two valences, viz; Mn^{3+} and Mn^{4+} ions. In the case of $La_{1-x}Sr_xMnO_3$, the samples with x=0.0-0.2 exhibit paramagnetic behavior and samples with x=0.3-0.5 exhibit ferromagnetic behavior at room temperature [Hemberger et al., 2002].

The M-H loops recorded at room temperature for the $La_{1-x}Sr_xMnO_3$ (x=0.3, 0.4 and 0.5) samples are shown in figure 2.39. All the prepared Sr doped lanthanum manganite samples exhibit hysteresis behavior, which indicates the ferromagnetic nature of the samples. This ferromagnetic behavior can be explained by the double exchange interaction taking place between the Mn^{3+} and Mn^{4+} ions along the Mn-O-Mn path. During the double exchange interaction, there is transfer of one electron from Mn^{3+} ($3d^4$, $t_{2g}^3 e_g^1$) orbital to an O (2p) orbital and another electron from the same orbital to Mn^{4+} ($3d^3$, $t_{2g}^3 e_g^0$) orbital on an adjacent ion. Hence, the e_g electrons move throughout the lattice and lead to the ferromagnetic behavior of the material [Jonker and Van Santen, 1950]. The magnetic parameters of the synthesized samples are given in table 2.21. It shows that the saturation magnetization decreases gradually for the increasing incorporation of Sr at the lattice site of La. This decrease in saturation magnetization may depend on the ratio Mn^{4+}/Mn^3. The slight variation in magnetic parameters may also depend on the inter-particle interactions, micro-strain and magneto crystalline anisotropy.

The saturation magnetization (M_s) values from VSM measurements, for all the doped lanthanum chromite and lanthanum manganite samples are given for comparison in table 3.7.

The room temperature M-H curves of lanthanum chromite samples confirm that all the samples exhibit antiferromagnetic behavior except the sample with x=0.12 in $(La_{0.8}Ca_{0.2})(Cr_{0.9-x}Co_{0.1}Fe_x)O_3$ series. From table 3.7, it is observed that, the saturation magnetization values for the lanthanum chromite samples vary from 1.14×10^{-3} emu g^{-1} to 38.12×10^{-3} emu g^{-1} and the sample $(La_{0.8}Ca_{0.2})(Cr_{0.78}Co_{0.1}Fe_{0.12})O_3$ has the maximum saturation magnetization value of 38.12×10^{-3} emu g^{-1}, which is the highest value when compared to the other chromite samples. So, the sample $(La_{0.8}Ca_{0.2})(Cr_{0.78}Co_{0.1}Fe_{0.12})O_3$ has good magnetic property than the other samples.

Table 3.7 *Saturation magnetization (M_s) values of doped lanthanum chromites and lanthanum manganites from VSM measurements*

Samples	Conc. (x)	Saturation magnetization $M_s \times 10^{-3}$ (emu g^{-1})
LCCCMn (300 K)	0.03	3.64
	0.06	7.90
	0.09	5.89
	0.12	8.76
LCCCFe (300 K)	0.03	11.20
	0.06	6.71
	0.09	10.47
	0.12	38.12
LCCCCu (300 K)	0.00	5.55
	0.03	2.43
	0.12	1.14
LCMO (20 K)	0.1	2189
	0.2	3015
	0.3	3354
	0.4	3438
	0.5	117
LCMO (300 K)	0.1	21.69
	0.2	58.35
	0.3	313.88
	0.4	347.30
	0.5	94.79
LSMO (300 K)	0.3	2118
	0.4	1692
	0.5	860

Conc. - Concentration
LCCCMn - $(La_{0.8}Ca_{0.2})(Cr_{0.9-x}Co_{0.1}Mn_x)O_3$
LCCCFe - $(La_{0.8}Ca_{0.2})(Cr_{0.9-x}Co_{0.1}Fe_x)O_3$
LCCCCu - $(La_{0.8}Ca_{0.2})(Cr_{0.9-x}Co_{0.1}Cu_x)O_3$
LCMO - $La_{1-x}Ca_xMnO_3$
LSMO - $La_{1-x}Sr_xMnO_3$

The M-H curves obtained at 300 K for $La_{1-x}Ca_xMnO_3$ indicate that all the samples (x=0.1, 0.2, 0.3, 0.4, 0.5) have paramagnetic nature. But, the M-H plots obtained at 300 K for $La_{1-x}Sr_xMnO_3$ confirm that all the samples (x=0.3, 0.4, 0.5) have ferromagnetic nature and the sample $La_{0.7}Sr_{0.3}MnO_3$ has good magnetic property than the other compositions.

The low temperature (20 K) M-H curves for $La_{1-x}Ca_xMnO_3$ confirm that the samples with x=0.1, 0.2, 0.3 and 0.4 have ferromagnetic nature and hence at 20 K, the sample $La_{0.6}Ca_{0.4}MnO_3$ has good magnetic property than the other samples. But, $La_{0.5}Ca_{0.5}MnO_3$ has the mixture of ferromagnetic and antiferromagnetic phases with low magnetization value of 0.117emu g^{-1}.

3.8 Charge density distribution analysis by maximum entropy method

The charge density distribution between the atoms in the unit cell of the doped lanthanum chromite and lanthanum manganite samples have been analyzed using the MEM [Collins, 1982] technique.

The MEM [Collins, 1982] computations for all the doped lanthanum chromite samples have been carried out considering $48\times64\times48$ pixels along the a, b and c-axes of the orthorhombic lattice.

3.8.1 (Co, Mn) doped (La, Ca) based chromites - $(La_{0.8}Ca_{0.2})(Cr_{0.9-x}Co_{0.1}Mn_x)O_3$

Figures 2.40 (a) - (d) show the three dimensional electron density distributions in the unit cell for all the synthesized samples, with same iso-surface level of 3.0 e/Å^3. The atomic positions of La, Cr and O atoms are clearly seen with charge distribution. The 3D electron density distribution in La-O2 bond for the co-doped samples are shown in figures 2.41 (a) - (d). From figures 2.41 (a) - (d), it is evident that the contour lines around O2 atom are denser than the La atom (electronegativity for O atom is 3.44 and that for La atom is 1.1) and there is no sharing of electrons between La and O2 atoms. Hence, La-O2 bond is more ionic. The 3D electron density distributions in Cr-O2 bond for the synthesized samples are shown in figures 2.41 (e) - (h). It is evident from figures 2.41 (e) - (h), that there is sharing of charges between Cr and O2 atoms and hence, the Cr-O2 bond is relatively covalent.

Figures 2.42 (a) and 2.43 (a) illustrate the 3D unit cells with (101) plane and (020) plane shaded. The 2D electron density distribution between La and O2 atoms are drawn in the range of 0-1.0 e/Å^3 with an interval 0.04 e/Å^3 on the (101) plane (figures 2.42 (b) - (e)) and between Cr and O2 atoms in the same range, on the (020) plane (figures 2.43 (b) - (e)). From the 2D electron density map for the La-O2 bond (figures 2.42 (b) - (e)), it is shown that the density of the contour lines are reduced from the boundary of the two atoms which confirms that there is no charge accumulation at the middle of the bond and hence the 2D map again confirms that, La-O2 bond is more ionic in nature. But, as far as the 2D electron density map for the Cr-O2 bond on the (020) plane is concerned, it is seen (figures 2.43 (b) - (e)) that there is an increase of charge accumulation in the bonding region between the Cr and O2 atoms and hence the Cr-O2 bond is more covalent in nature [Montross, 1997].

To explain the bonding features in a detailed manner, one dimensional electron density profiles for the bonds La-O2 and Cr-O2 are drawn as shown in figures 2.44 and 2.45. The bond length and the mid bond density values of La-O2 and Cr-O2 bonds for $(La_{0.8}Ca_{0.2})(Cr_{0.9-x}Co_{0.1}Mn_x)O_3$, (x=0.03, 0.06, 0.09 and 0.12) samples are presented in table 2.22. For the La-O2 bond, the mid bond density ranges from 0.147 e/$Å^3$ to 0.475 e/$Å^3$ and it confirms that the La-O2 bond is relatively ionic in nature. But, for the Cr-O2 bond, the mid bond density ranges from 0.520 e/$Å^3$ to 0.601 e/$Å^3$ which confirms that the Cr-O2 bond is more covalent in nature. The variations in bond length for La-O2 and Cr-O2 bonds, for all compositions of Mn (x=0.03, 0.06, 0.09 and 0.12) exactly follow the trend in the XRD results.

The oxygen bonds O1-O2 and O2(A)-O2(B) in the 3D unit cell of $(La_{0.8}Ca_{0.2})(Cr_{0.87}Co_{0.1}Mn_{0.03})O_3$ are shown in figure 2.46. The one dimensional electron density profiles for the oxygen bonds O1-O2 and O2(A)-O2(B) are also analyzed which are shown in figures 2.47 and 2.48. Table 2.23 gives the bond length and mid bond density values of the O1-O2 and O2(A)-O2(B) bonds. The mid bond density for the O1-O2 bond ranges between 0.148 e/$Å^3$ and 0.305 e/$Å^3$ and these values confirm that the bond O1-O2 is more ionic. For the O2(A)-O2(B) bond, mid bond density ranges between 0.444 e/$Å^3$ to 0.567 e/$Å^3$ and these values confirm that the bond O2(A)-O2(B) is less ionic.

3.8.2 (Co, Fe) doped (La, Ca) based chromites - $(La_{0.8}Ca_{0.2})(Cr_{0.9-x}Co_{0.1}Fe_x)O_3$

The 3D electron density distributions in the unit cell with similar iso-surface level of 3.0 e/$Å^3$ for all the synthesized samples are shown using ball and stick model of structure and are presented in figures 2.49 (a) - (d). The 3D plots shown in figures 2.49 (a) - (d), clearly indicate the positions of the La, Cr and O atoms and confirms the orthorhombic structure of the samples. The 3D electron density distributions for the La-O2 bond for all Fe concentrations (x=0.03, 0.06, 0.09 and 0.12) are shown in figures 2.50 (a) - (d) and visualize that, there is no sharing of electrons between the La and O2 atoms and confirms the ionic nature of the La-O2 bond. The 3D electron density distributions of the Cr-O2 bond are shown in figures 2.50 (e) - (h), which clearly visualizes the electron clouds around the Cr and O2 atoms. Again, figures 2.50 (e) - (h) reveal that, the electrons are shared between the Cr and O2 atoms and confirm that the Cr-O2 bond is more covalent [Montross, 1997].

Figures 2.51 (a) and 2.52 (a) show the three dimensional unit cells with the (101) and (020) planes shaded. The 2D electron density distributions for the La-O2 and Cr-O2 bonds are drawn in the range 0-1.0 e/Å^3 with an interval 0.04 e/Å^3 on the (101) plane and (020) plane and are shown in figures 2.51 (b) - (e) and 2.52 (b) - (e) respectively. The 2D electron density maps for the La-O2 bond (figures 2.51 (b) - (e)) show that there is no sharing of electrons between the La and O2 atoms, which again confirms that, the bond La-O2 is ionic in nature. The 2D electron density maps for the Cr-O2 bond (figures 2.52 (b) - (e)) reveal that there is sharing of electrons between the Cr and O2 atoms, which again confirm that the Cr-O2 bond is more covalent in nature.

To explain the above facts in detail, the one dimensional electron density profiles for the two bonds La-O2 and Cr-O2 are drawn and shown in figures 2.53 and 2.54. The bond lengths and mid bond density values are tabulated in table 2.24. For the La-O2 bond, the mid bond density ranges from 0.368 e/Å^3 to 0.802 e/Å^3. These mid bond values confirm that, the incorporation of Fe at the lattice site of Cr decreases the ionic behavior of the La-O2 bond. For the Cr-O2 bond, the mid bond density ranges from 0.357 e/Å^3 to 0.522 e/Å^3, which lead to the decrease in ionic behavior of the Cr-O2 bond. The bond length variation in the La-O2 and Cr-O2 bonds for all compositions of Fe (x=0.03, 0.06, 0.09 and 0.12) exactly follow the trend in the XRD results. For the sample with x=0.12, the mid bond electron density values for the La-O2 and Cr-O2 bonds are 0.409 e/Å^3 and 0.441 e/Å^3 respectively. These mid bond values reveal that, there is a significant increase of charge accumulation at the mid of the bonds which enhance the electrical conduction property of the material. The 1D electron density profiles for the O1-O2 bond for the synthesized samples are shown in figure 2.55. The mid bond density values for the O1-O2 bond given in table 2.24 confirm that the O1-O2 bond is ionic with partly covalent in nature.

3.8.3 (Co, Cu) doped (La, Ca) based chromites - $(La_{0.8}Ca_{0.2})(Cr_{0.9-x}Co_{0.1}Cu_x)O_3$

For the synthesized (Co, Cu) doped (La, Ca) based chromites, the 3D electron density distributions in the unit cell are constructed with the same iso-surface level of 3.0 e/Å^3 and are presented in figures 2.56 (a) - (c). To analyze the La-O and Cr-O bonding behavior, the electron density distributions on two different crystallographic planes (101) and (020) have been examined. Figure 2.57 (a) illustrates the 3D unit cell with the (101) plane shaded. The 2D electron density contour maps for the La-O2 and Cr-O2 bonds are drawn in the range 0-1.0 e/Å^3 with an interval 0.04 e/Å^3. Figures 2.57 (b) - (d) show the two dimensional electron density contour maps corresponding to the (101) plane for

$(La_{0.8}Ca_{0.2})(Cr_{0.9-x}Co_{0.1}Cu_x)O_3$ (x=0.00, 0.03 and 0.12) samples. These 2D maps visualize that there is not much charge sharing between the La and O2 atoms and confirm the ionic nature of the La-O2 bond. Figure 2.58 (a) represents the 3D unit cell with the (020) plane shaded and figures 2.58 (b) - (d) show the two dimensional electron density contour maps corresponding to the (020) plane with Cr-O2 bonding. These 2D maps show that there is an increase in bond charges between the Cr and O2 atoms and confirms the covalent nature of the Cr-O2 bond.

To quantify the above results, one dimensional electron density profiles for La-O2 and Cr-O2 bonds are drawn and are shown in figures 2.59 and 2.60. Table 2.25 gives the bond lengths and the mid bond density values of the La-O2 and Cr-O2 bonds. The mid bond electron density values of the La-O2 bond confirm that, Cu addition enhances the ionic nature between the La and O2 ions. The mid bond electron density values of the Cr-O2 bond confirm that, Cu addition reduces the covalent nature between the Cr and O2 ions. This reduction in covalent character existing in the Cr-O2 bond may be attributed to the antiferromagnetic behavior of the synthesized samples [Rashad and El-Sheikh, 2011]. Figure 2.61 shows the 1D profile for the oxygen bond O1-O2. The mid bond density values of the O1-O2 bond confirm that the bond O1-O2 is ionic in nature.

The CrO_6 octahedron constructed using VESTA [Momma and Izumi, 2008] for the sample $(La_{0.8}Ca_{0.2})(Cr_{0.87}Co_{0.1}Cu_{0.03})O_3$ is shown in figure 2.62. Table 2.26 gives the bond lengths of O1-O1, O2(2)-O2(4) and O2(1)-O2(3) bonds for all the Cu concentrations of the synthesized samples. In perovskite structure (ABO_3), the octahedral (BO_6) tilting induces spin canting which may be attributed to the ferromagnetic behavior of the material. An octahedral tilt causes an expansion along the bond length between the two diagonally opposite oxygen (O2(2)-O2(4)) atoms and a contraction along the bond length between another two diagonally opposite oxygen (O2(1)-O2(3)) atoms [Shukla et al., 2009]. Table 2.26 shows that, the O2(2)-O2(4) bond expands and O2(1)-O2(3) bond contracts, with increasing incorporation of Cu at the lattice site of Cr. Moreover, the bond length distortion parameters obtained from VESTA [Momma and Izumi, 2008] for x=0.00, 0.03 and 0.12 samples are 0.035, 0.023 and 0.005. Since these parameters are very low, the CrO_6 octahedron distortion can be considered as minor distortion and there is no spin canting due to Cr^{3+} ions and hence the prepared samples exhibit antiferromagnetism.

The nature of bond and the mid bond density values of La-O2, Cr-O2 and O1-O2 bonds for the lanthanum chromite samples are presented in table 3.8. For the $(La_{0.8}Ca_{0.2})(Cr_{0.9-x}Co_{0.1}Mn_x)O_3$, (x=0.03, 0.06, 0.09 and 0.12) samples, the mid bond electron density values for La-O2 and Cr-O2 bonds confirm that the bond La-O2 has ionic nature and Cr-O2 has covalent nature. For $(La_{0.8}Ca_{0.2})(Cr_{0.9-x}Co_{0.1}Fe_x)O_3$ samples, the mid bond density ranges between 0.368 e/Å^3 and 0.802 e/Å^3 which indicate that the ionic nature of the La-O2 bond decreases for the incorporation of Fe at the Cr site of the lattice. The mid bond density for the Cr-O2 bond ranges between 0.357 e/Å^3 and 0.552 e/Å^3 which confirms that covalency increases by the doping of Fe. Table 3.8 shows that the ionic nature of La-O2 bond increases and covalent nature of Cr-O2 bond decreases by the substitution of Cu at the Cr site of the lattice for $(La_{0.8}Ca_{0.2})(Cr_{0.9-x}Co_{0.1}Cu_x)O_3$ samples.

Table 3.8 *The mid bond density values and the nature of bond for La-O2, Cr-O2 and O1-O2 bonds for the lanthanum chromite samples.*

Samples	Conc. (x)	La-O2		Cr-O2		O1-O2	
		Mid bond density (e/Å^3)	Relative nature of bond	Mid bond density (e/Å^3)	Relative nature of bond	Mid bond density (e/Å^3)	Relative nature of bond
	0.03	0.328	ionic	0.601	covalent	0.154	ionic
	0.06	0.475	ionic	0.545	covalent	0.148	ionic
LCCCMn	0.09	0.147	ionic	0.520	covalent	0.305	ionic
	0.12	0.461	ionic	0.550	covalent	0.244	ionic
	0.03	0.368	ionic	0.421	covalent	0.248	ionic
LCCCFe	0.06	0.802	Less ionic	0.522	covalent	0.406	ionic
	0.09	0.455	ionic	0.357	Less covalent	0.152	ionic
	0.12	0.409	ionic	0.441	covalent	0.239	ionic
	0.00	0.624	Less ionic	0.504	covalent	0.401	ionic
LCCCCu	0.03	0.385	ionic	0.391	Less covalent	0.259	ionic
	0.12	0.303	ionic	0.429	covalent	0.381	ionic

Conc. - Concentration
LCCCMn - $(La_{0.8}Ca_{0.2})(Cr_{0.9-x}Co_{0.1}Mn_x)O_3$
LCCCFe - $(La_{0.8}Ca_{0.2})(Cr_{0.9-x}Co_{0.1}Fe_x)O_3$
LCCCCu - $(La_{0.8}Ca_{0.2})(Cr_{0.9-x}Co_{0.1}Cu_x)O_3$

3.8.4 $La_{1-x}Ca_xMnO_3$ manganites

The three dimensional electron density distribution in the unit cell of Ca doped lanthanum manganite samples with similar iso-surface level of 3.0 e/Å^3 are shown in figures 2.63 (a) - (e). The positions of the La, Mn and O atoms have been clearly visualized in the 3D unit cell. Figures 2.64 (a) and 2.65 (a) represent the 3D unit cells with the (101) plane and (020) plane shaded. The 2D electron density distribution maps for the La-O2 bond are drawn in the range of 0-1.0 e/Å^3 with an interval of 0.04 e/Å^3 on the (101) plane (figures 2.64 (b) - (f)) and for the Mn-O2 bond in the same range on the (020) plane (figures 2.65 (b) - (f)). The contour lines around the La and O2 atoms in figures 2.64 (b) - (f) indicate that there is no sharing of charges in the bonding region (electronegativity for the La and O atoms are 1.1 and 3.44 respectively) which confirms, the La-O2 bond is more ionic with partial covalent character. The 2D maps (figures 2.65 (b) – (f)) for the Mn-O2 bond on the (020) plane show that there is sharing of charges along the bonding region between the Mn and O2 atoms. It confirms that the Mn-O2 bond is more covalent [Alonso et al., 2000].

The La-O2 and Mn-O (Mn-O2 along a-axis, Mn-O1 along b-axis and Mn-O2 along c-axis) bonds can be quantitatively analyzed by the 1D electron density profiles. Figures 2.66 and 2.67 illustrate the 1D electron density profiles of the La-O2 and Mn-O bonds. Table 2.27 gives the bond length and mid bond density values for La-O2 bond. The mid bond density values for La-O2 bond range between 0.318 e/Å^3 and 0.667 e/Å^3 and confirm that the La-O2 bond is more ionic with partial covalent character.

Table 2.28 gives the bond length and mid bond density values of all Mn-O bonds (Mn-O2 (a-axis), Mn-O1 (b-axis) and Mn-O2 (c-axis)) for La$_{1-x}$Ca$_x$MnO$_3$, (x=0.1, 0.2, 0.3, 0.4, 0.5). The mid bond electron density values for Mn-O2 bonds (along a and b-axes) reveal that the covalency increases by the incorporation of Ca at the lattice site of La. It is also noticed that, for the sample with x=0.4, the Mn-O2 bonds are more covalent than the other samples (x=0.1, 0.2, 0.3 and 0.5). The mid bond electron density values of Mn-O1 bond range from 0.514 e/Å^3 to 1.025 e/Å^3 and the bond is more covalent for the sample with x=0.4. For the sample with x=0.5, the ionic nature of the La-O2 bond as well as the covalent nature of all Mn-O bonds (Mn-O2 along a-axis, Mn-O1 along b-axis and Mn-O2 along c-axis) decreases. This attitude can also be noticed in the M-H curves drawn at 300 K (figure 2.38), where the paramagnetic behavior decreases at x=0.5. With increasing incorporation of Ca, the bond lengths for all bonds (La-O2, Mn-O bonds) decrease. The bond length variation for various doping concentrations of calcium exactly follows the trend in the XRD results.

Table 3.9 gives the nature of the bond and the mid bond density values for the La-O2 and Mn-O (Mn-O2 along a-axis, Mn-O1 along b-axis and Mn-O2 along c-axis) bonds for the calcium doped lanthanum manganite samples. It confirms that, for Ca doped manganite samples, the ionic nature between the La and O2 atoms decreases and the covalent nature between the Mn and O2 atoms increases by the incorporation of Ca at the La site of the lattice. The sample La$_{0.6}$Ca$_{0.4}$MnO$_3$ has the highest mid bond electron density value of

1.025 e/Å^3 and confirms that, this sample has more conducting property than other samples (x=0.1, 0.2, 0.3 and 0.5).

Table 3.9 *The mid bond density values and the nature of bond for the La-O2 and Mn-O (Mn-O2 along a-axis, Mn-O1 along b-axis and Mn-O2 along c-axis) bonds for La$_{1-x}$Ca$_x$MnO$_3$.*

	Bonds							
	La-O2		**Mn-O2(a-axis)**		**Mn-O1(b-axis)**		**Mn-O2(c-axis)**	
Conc. (x)	Mid bond density (e/Å^3)	Relative nature of bond	Mid bond density (e/Å^3)	Relative nature of bond	Mid bond density (e/Å^3)	Relative nature of bond	Mid bond density (e/Å^3)	Relative nature of bond
0.1	0.318	ionic	0.268	Less covalent	0.992	covalent	0.389	Less covalent
0.2	0.667	Less ionic	0.354	Less covalent	0.784	covalent	0.588	covalent
0.3	0.436	Ionic	0.468	covalent	0.514	covalent	0.489	covalent
0.4	0.397	ionic	0.539	covalent	1.025	covalent	0.676	covalent
0.5	0.452	ionic	0.423	covalent	0.589	covalent	0.531	covalent

3.8.5 La$_{1-x}$Sr$_x$MnO$_3$ manganites

MEM [Collins, 1982] computations for the prepared La$_{1-x}$Sr$_x$MnO$_3$, (x=0.3, 0.4 and 0.5) samples have been carried out considering 54×54×144 pixels along a, b and c-axes of the rhombohedral lattice. The 3D charge density distributions of the synthesized samples are constructed with an iso-surface level of 1.5 e/Å^3 and are shown in figures 2.68 (a) - (c). The 3D unit cells clearly visualize the positions of the La, Mn and O atoms. To analyze electron density distributions along the La-O and Mn-O bonds, the 2D contour maps for two different crystalline planes (004) and (012) have been drawn in the range of 0-1.0 e/Å^3 with an interval 0.05 e/Å^3. Figures 2.69 (a) and 2.70 (a) show the three dimensional unit cells with the (004) and (012) planes shaded. The 2D maps in figures 2.69 (b) - (d) show that, there is no charge sharing between the La and O atoms. It confirms that the La-O bond is more ionic in nature. Figures 2.70 (b) - (d) show the 2D contour maps for the (012) plane of the unit cell for all the Sr doped samples. The 2D maps (figures 2.70 (b) - (d)) for the (012) plane show that the charges are shared between the Mn and O atoms along the bonding region between them and confirm that the Mn-O bond is more covalent in nature.

To explain the La-O and Mn-O bonding features in a quantitative manner, one dimensional profiles for the La-O and Mn-O bonds are drawn and are shown in figures 2.71 and 2.72 respectively. The bond lengths and mid bond electron density values are tabulated in table 2.29. The La-O, Mn-O and O-O bond lengths are found to be decreasing with the incorporation of Sr at the lattice site of La, which are consistent with the XRD results. The mid bond electron density values for the La-O bond indicate that, the ionic nature is enhanced for the increasing incorporation of Sr. The mid bond electron density values given in table 2.29, for the Mn-O bond again confirm that the Mn-O bond is more covalent in nature. The covalency existing in the Mn-O bond may be attributed to the ferromagnetic behavior of the synthesized samples. Figure 2.73 shows the one dimensional profile for the O-O bond. The mid bond electron density values of the O-O bond confirm that the bond is more ionic with less covalent behavior.

The nature of bond and the mid bond density values of the La-O, Mn-O and O-O bonds for the strontium doped lanthanum manganite samples have been presented in table 3.10. For Sr doped manganite samples, the ionic nature of La-O bond increases and the covalent nature of the Mn-O bond decreases by the substitution of the Sr at the La site of the lattice. The mid bond electron density value for the $La_{0.7}Sr_{0.3}MnO_3$ sample (0.530 e/$Å^3$) confirms more conduction, as compared to the other samples (x=0.4, 0.5).

Table 3.10 *The mid bond density values and the nature of the bond for the La-O, Mn-O and O-O bonds for $La_{1-x}Sr_xMnO_3$.*

Conc. (x)	Bonds					
	La-O		Mn-O		O-O	
	Mid bond density (e/$Å^3$)	Relative nature of bond	Mid bond density (e/$Å^3$)	Relative nature of bond	Mid bond density (e/$Å^3$)	Relative nature of bond
0.3	0.507	ionic	0.530	covalent	0.283	ionic
0.4	0.451	ionic	0.440	covalent	0.347	ionic
0.5	0.214	ionic	0.456	covalent	0.356	ionic

References

[1]   Alonso J. A., Martı'nez-Lope M. J, and Casais M. T., Inorg. Chem., 39, 917 (2000). https://doi.org/10.1021/ic990921e

[2]   Alvarez G.A., Wang X.L., Peleckis G., Dou S.X., Zhu J.G. and Lin Z.W., J. Applied Phys. 103, 07B916 1 (2008)

[3]   Brajesh T., Dixit A., Naik R., Lawes G. and Rama Chandra Rao M.S., Materials Research Express, 2, 1 (2015)

[4]   Chaudhary T, Khamar J, Chaudhary P, Chaudhary V, Barot M and Jyoti Sen D., J. Drug Discovery and Therapeutics, 3, 09 (2015)

[5]   Choudhry Q., Azhar Khan M., Nasar G., Mahmood A., Imran Shakir M. and Farooq Warsi M. (2015) doi: http://dx.doi.org/10.1016/j.jmmm.2015.05.040. https://doi.org/10.1016/j.jmmm.2015.05.040

[6]   Coey J M D, Viret M, von Molnair S, Adv. Phys., 48, 167 (1999). https://doi.org/10.1080/000187399243455

[7]   Collins D. M., Nature 298, 49 (1982). https://doi.org/10.1038/298049a0

[8]   Culllity B.D, Stock S.R. Elements of X-ray Diffraction, third ed. Prentice Hall, New Jersy, 2001

[9]   Elemans J. B. A. A., Larr. B. V., Van Der Veen K. R. and Loopstra B. O., J. of solid state chem., 3, 238 (1971). https://doi.org/10.1016/0022-4596(71)90034-X

[10]   Ewe L.S, Ramli R, Lim K.P. and Abd-Shukor R, Sains Malaysiana., 41, 761 (2012)

[11]   Gonjal J. P., Schmidt R., Romero J. J., Amador D. U. and Moran E., Inorg. Chem., 52, 313 (2013). https://doi.org/10.1021/ic302000j

[12]   Gupta R.K and Whang C. M, J. of Physics: Condensed Matter, 19, 1 (2007). https://doi.org/10.1088/0953-8984/19/19/196209

[13]   Hammouche A., Siebert E. and Hammou A., Mat. Res. Bull., 24, 367 (1989). https://doi.org/10.1016/0025-5408(89)90223-7

[14]   Hemberger J., Krimmel A., Kurz T., Krug von Nidda H. A., Yu V. Ivanov, Mukhin A. A., Balbashov A. M. and Loidl, A., Phys. Rev. B, 66, 1 (2002)

[15]   Israel S., Saravana kumar S., Renuretson R., Sheeba R. A. J. R and Saravanan R., Bull. Mater. Sci., 35, 111 (2012)

[16]   JCPDS PDF # 33-0701

[17]   JCPDS PDF # 72-0842

[18]   JCPDS PDF # 47-0444

[19]   Jonker G. H, Van Santen J. H., Physica XVI, 3, 337 (1950)

[20]   Jung J. H, Kim K. H, and Noh T. W. Phys. Rev. B., 57, 043 (1998)

[21]   Lira-Herna'ndez Ivan A., Felix Sa'nchez-De Jesu's, Claudia A.Corte's-Escobedo and Ana M. Boları'n-Miro´, J. Am. Ceram. Soc., 93, 3474 (2010)

[22]   Loa I, Adler P, Grzechnik A, Syassen K, Schwarz U, Hanfland M, Rozenberg G. Kh, Gorodetsky P and Pasternak M. P., Phys. Rev. Lett., 87, 1 (2001). https://doi.org/10.1103/PhysRevLett.87.125501

[23]   Mitra C., Raychaudhuri P., John J., Dhar S. K., Nigam A.K., and Pinto R, J. Appl. Phys., 89, 524 (2001). https://doi.org/10.1063/1.1331648

[24]   Momma K, Izumi F, VESTA: a three-dimensional visualization system for electronic and structural analysis. J. Appl. Crystallogr., 41, 653 (2008). https://doi.org/10.1107/S0021889808012016

[25]   Montross C. S., J. European Ceramic Society, 18, 353 (1997). https://doi.org/10.1016/S0955-2219(97)00143-X

[26]   Naseem S, Khan W, Saad A.A, Shoeb M., Ahmed H, Husain S and Naqvi A.H., AIP Conference Proceedings 1591, 259 (2014). https://doi.org/10.1063/1.4872565

[27]   Ong K.P., Peter Blaha and Ping Wu, Phys. Rev. B., 77, 073102 1 (2008)

[28]   Petricek V, Dusek M, Palatinus L, Kristallogr Z, Crystallographic Computing System JANA2006: General features, 229, 345 (2014)

[29]   Qing-Gong S., Lingling S., Hui Z., Tong W. and Jianhai K., Adv. Mate. Res. 622-623, 734 (2013)

[30]   Ramirez A. P. J. Phys.: Condens. Matter., 9, 8171 (1997). https://doi.org/10.1088/0953-8984/9/39/005

[31]   Rashad M. M., El-Sheikh S. M., Mater. Res. Bull., 46, 469 (2011). https://doi.org/10.1016/j.materresbull.2010.10.016

[32]   Rietveld H.M., J. Appl. Crystallogr., 2, 65 (1969). https://doi.org/10.1107/S0021889869006558

[33]   Ruben A. D., Fugio I., Superfast program PRIMA for the Maximum Entropy Method, Advanced Materials Laboratory, National Institute for Material Science, Ibaraki, Japan (2004), 3050044

[34]   Saravanan R., Saravanakumar S., Lavanya S., Physica B, 405, 3700 (2010). https://doi.org/10.1016/j.physb.2010.05.069

[35]   Saravanan R., GRAIN software, Private Communication, (2008)

[36] Shannon R.D., Acta Cryst. A, 32, 751 (1976).
      https://doi.org/10.1107/S0567739476001551

[37] Shukla R., Manjanna J., Bera A.K, Yusuf S. M., and Tyagi A. K., Inorg. Chem.,
      48, 11691 (2009). https://doi.org/10.1021/ic901735d

[38] Sultan K, Ikram M. Adv. Mater. Lett. 6, (749 (2015)

[39] Szytuła A., Acta Physica Polonica A, 118, 303 (2010).
      https://doi.org/10.12693/APhysPolA.118.303

[40] Thiesse A, Beyreuther E, Werner R, Koelle D, Kleiner R, En L.M, J. Phys. Chem.
      Solids., 80, 26 (2015). https://doi.org/10.1016/j.jpcs.2014.12.014

[41] Terashita H., Cezar J.C., Ardito F.M., Bufaical L.F. and Granado E., Phys. Rev. B
      85, 104401 (2012). https://doi.crg/10.1103/PhysRevB.85.104401

[42] Tribedi Bora and Ravi S., J. applied phys., 114, 033906 (2013).
      https://doi.org/10.1063/1.4813516

[43] Tyson T.A., Qian Q., Kao C., Rueff J.P., F. M. F. de Groot F. M., Croft M.,
      Cheong S.W., Greenblatt M., Subramanian M. A., Phys. Rev. B., 60, 4655 (1999)

[44] Wood D. L, Tauc J, Phys Rev B., 5, 3144 (1972).
      https://doi.org/10.1103/PhysRevB.5.3144

Chapter 4

Conclusion

Abstract

Chapter 4 presents the conclusion of the results of the reported work.

Keywords

Lanthanam Chromite, Solid State Reaction, Structure, Surface Morphology, Elemental Composition, Optical, Magnetic, Charge Density

Contents

4.1 Conclusion

In the present work, the following samples having manganite structure have been synthesized using the high temperature solid state reaction method.

1. (Co, Mn) doped (La, Ca) based chromites - $(La_{0.8}Ca_{0.2})(Cr_{0.9-x}Co_{0.1}Mn_x)O_3$

2. (Co, Fe) doped (La, Ca) based chromites - $(La_{0.8}Ca_{0.2})(Cr_{0.9-x}Co_{0.1}Fe_x)O_3$

3. (Co, Cu) doped (La, Ca) based chromites - $(La_{0.8}Ca_{0.2})(Cr_{0.9-x}Co_{0.1}Cu_x)O_3$

4. $La_{1-x}Ca_xMnO_3$ manganites

5. $La_{1-x}Sr_xMnO_3$ manganites

The above mentioned doped lanthanum chromite and lanthanum manganite samples were characterized using powder X-ray diffraction (XRD), scanning electron microscopy (SEM), energy dispersive X-ray spectroscopy (EDS), UV-vis absorption spectroscopy

Solid Oxide Fuel Cell (SOFC) Materials

(UV-vis) and vibrating sample magnetometry (VSM). The spatial charge density distributions as a whole in the unit cell for all the synthesized samples were analyzed thoroughly using maximum entropy method (MEM). In this section, the investigations from the above characterization techniques and the conclusions are presented.

a)   Structural properties

All the doped lanthanum chromite and lanthanum manganite samples have been characterized using powder X-ray diffraction for their structural properties. The observed XRD data sets of all the doped lanthanum chromite and lanthanum manganite samples have been subjected to powder profile refinement, to obtain detailed structural information.

The powder X-ray diffraction patterns of all the synthesized (Co, Mn), (Co, Fe) and (Co, Cu) doped (La, Ca) based chromites confirm the orthorhombic phase of the samples.

The powder X-ray diffraction patterns of (Co, Mn) doped (La, Ca) based chromite samples $(La_{0.8}Ca_{0.2})(Cr_{0.9-x}Co_{0.1}Mn_x)O_3$, (x=0.03, 0.06, 0.09 and 0.12), show that the Bragg peaks shift towards the right and to the left alternatively for the increasing Mn incorporation in the Cr site of the lattice. This shifting of the Bragg peaks indicates that there is no systematic variation in the unit cell volume for the synthesized co-doped samples. The refined profiles of $(La_{0.8}Ca_{0.2})(Cr_{0.9-x}Co_{0.1}Mn_x)O_3$, (x=0.03, 0.06, 0.09 and 0.12) show that there is a good match between the observed and the calculated XRD patterns.

The powder X-ray diffraction patterns of (Co, Fe) doped (La, Ca) based chromite samples $(La_{0.8}Ca_{0.2})(Cr_{0.9-x}Co_{0.1}Fe_x)O_3$, (x=0.03, 0.06, 0.09 and 0.12) indicate that the peaks corresponding to the Fe composition of x=0.06 and 0.09 shift towards the lower angle side with respect to the x=0.03 composition. But for x=0.12 composition, the XRD peaks shift slightly towards the higher angle side with respect to the x=0.03 composition. The refined structural parameters follow the trend observed in the Bragg peak shifting.

The powder X-ray diffraction patterns of (Co, Cu) doped (La, Ca) based chromite samples $(La_{0.8}Ca_{0.2})(Cr_{0.9-x}Co_{0.1}Cu_x)O_3$, (x=0.00, 0.03 and 0.12) show that the Bragg peaks shift towards the higher angle side with respect to the undoped (x=0.00) sample for increasing concentration of Cu. The refined structural parameters confirm that, with the incorporation of Cu at the Cr site of the lattice, the lattice parameters and unit cell volume decrease.

All the synthesized doped lanthanum chromite samples have orthorhombic crystal structure. The orthorhombic unit cell has corner-linked octahedra CrO_6. The centre of the octahedron is occupied by centrosymmetric chromium (Cr) ions and the oxygen ions (O) are at the vertices of the octahedron. Lanthanum ions (La) occupy the space between the octahedra.

The average grain size of the synthesized doped lanthanum chromite samples ranges from 13 nm to 31 nm.

The powder X-ray diffraction patterns of $La_{1-x}Ca_xMnO_3$ (x=0.1, 0.2, 0.3, 0.4 and 0.5) confirm the orthorhombic structure of the samples. The powder XRD patterns of these manganite samples show a right shift of the diffraction peaks with respect to increasing Ca doping concentration. The refined structural parameters indicate the shrinkage in the unit cell for the increasing incorporation of calcium. The average grain size of the synthesized manganite samples $La_{1-x}Ca_xMnO_3$, (x=0.1, 0.2, 0.3, 0.4 and 0.5) ranges from 35 nm to 63 nm.

The powder X-ray diffraction patterns of $La_{1-x}Sr_xMnO_3$, (x=0.3, 0.4 and 0.5) confirm the rhombohedral structure of these samples. The powder XRD patterns of strontium doped lanthanum manganite samples show that the Bragg peaks shift towards the higher angle (2θ) side for increasing incorporation of Sr. A shrinkage in the unit cell is observed from the refined structural parameters. The average grain size of the synthesized $La_{1-x}Sr_xMnO_3$, (x=0.3, 0.4 and 0.5) manganite samples ranges from 10 nm to 15 nm.

b)   Surface morphology and elemental compositions

The surface morphology and the elemental composition of all the doped lanthanum chromite and lanthanum manganite samples have been analyzed through the SEM images and the EDS spectra.

The SEM images of (Co, Mn) doped (La, Ca) based chromite samples $(La_{0.8}Ca_{0.2})(Cr_{0.9-x}Co_{0.1}Mn_x)O_3$, (x=0.03, 0.06, 0.09 and 0.12) reveal that the particles appear in spherical shape and are distributed without much agglomeration. In the case of (Co, Fe) doped (La, Ca) based chromite samples $(La_{0.8}Ca_{0.2})(Cr_{0.9-x}Co_{0.1}Fe_x)O_3$, (x=0.03, 0.06, 0.09 and 0.12), the particles are in spherical shape with different sizes and are heterogeneously distributed. The SEM images of (Co, Cu) doped (La, Ca) based chromite samples $(La_{0.8}Ca_{0.2})(Cr_{0.9-x}Co_{0.1}Cu_x)O_3$, (x=0.00 and 0.03) show the presence of tiny individual particles of different sizes. But, for $(La_{0.8}Ca_{0.2})(Cr_{0.9-x}Co_{0.1}Cu_x)O_3$, (x=0.12) sample, particles of different sizes have been distributed heterogeneously with voids. The average particle size for the synthesized lanthanum chromite samples ranges from 0.23 µm to 0.66 µm.

The SEM images of the synthesized calcium doped lanthanum manganite samples $La_{1-x}Ca_xMnO_3$ (x=0.1, 0.2, 0.3, 0.4 and 0.5) visualize the particles in polygonal form with different sizes. These particles are highly aggregated with no voids. In the case of strontium doped lanthanum manganite samples $La_{1-x}Sr_xMnO_3$, (x=0.3, 0.4 and 0.5), aggregated particles of polygonal shape with no voids are observed from the SEM images. The particles have clear grain boundary between them. The SEM images of both Ca and Sr doped lanthanum manganite samples reveal that, there is a significant change in the particle size for the different doping compositions.

The EDS spectra for all the synthesized doped lanthanum chromite and lanthanum manganite samples confirm that, no other impurities are present in the synthesized samples.

c)   Optical studies

The UV-visible absorption spectra of all the doped lanthanum chromite and lanthanum manganite samples were recorded in the range between 200 nm and 2000 nm. Using Tauc plot, the optical band gap of the synthesized materials has been estimated.

For $(La_{0.8}Ca_{0.2})(Cr_{0.9-x}Co_{0.1}Mn_x)O_3$ samples, the optical band gap values range from 2.270 eV to 2.464 eV. For $(La_{0.8}Ca_{0.2})(Cr_{0.9-x}Co_{0.1}Fe_x)O_3$ samples, the optical band gap values range from 2.135 eV to 2.405 eV. For this case, as the Fe doping concentration increases, the optical band gap (E_g) values are found to be decreasing. For (Co, Cu) doped (La, Ca) based chromite samples $(La_{0.8}Ca_{0.2})(Cr_{0.9-x}Co_{0.1}Cu_x)O_3$, the E_g values range from 1.859 eV to 2.448 eV. For this case also, the optical band gap (E_g) values are found to be decreasing with the increasing concentration of Cu at the lattice site of Cr.

For calcium doped lanthanum manganites $La_{1-x}Ca_xMnO_3$, the optical band gap values range between 1.411 eV and 1.730 eV. The E_g values are decreasing for the increasing doping concentration of Ca. For $La_{1-x}Sr_xMnO_3$, the optical band gap values range from 2.442 eV to 2.487 eV.

d)   Magnetic properties

The room temperature M-H curves of all the doped lanthanum chromite and lanthanum manganite samples were recorded using a vibrating sample magnetometer (VSM). Low temperature (20 K) M-H curves for calcium doped lanthanum manganite samples were also recorded using a VSM.

M-H curves of $(La_{0.8}Ca_{0.2})(Cr_{0.9-x}Co_{0.1}Mn_x)O_3$, (x=0.03, 0.06, 0.09 and 0.12) chromite samples confirm that all these samples exhibit predominant antiferromagnetic ordering due to Cr^{3+} spins. M-H curves of $(La_{0.8}Ca_{0.2})(Cr_{0.9-x}Co_{0.1}Fe_x)O_3$, (x=0.03, 0.06 and 0.09) chromite samples exhibit minor hysteresis loop with low saturation magnetization. But, the sample with x=0.12 exhibits large hysteresis with high saturation magnetization which is due to considerable increase of Fe concentration and associated uncompensated canted FM. The M-H curves of $(La_{0.8}Ca_{0.2})(Cr_{0.9-x}Co_{0.1}Cu_x)O_3$, (x=0.00, 0.03 and 0.12) samples do not show any hysteresis and they exhibit predominant antiferromagnetic ordering due to Cr^{3+} spins.

The M-H loops recorded at 20 K of $La_{1-x}Ca_xMnO_3$, (x=0.1, 0.2, 0.3, 0.4 and 0.5) manganite samples exhibit ferromagnetic behavior which is due to the double exchange interaction taking place between the Mn^{3+} and Mn^{4+} ions. But, the M-H loops recorded at 300 K for the $La_{1-x}Ca_xMnO_3$ (x=0.1, 0.2, 0.3, 0.4 and 0.5) manganite samples exhibit paramagnetic behavior. The reason for paramagnetic behavior is that the Mn spin fluctuations are so high that an applied field cannot align the spins at room temperature.

The M-H loops at 300 K of the $La_{1-x}Sr_xMnO_3$, (x=0.3, 0.4 and 0.5) manganite samples exhibit ferromagnetism which is due to the double exchange interaction taking place between the Mn^{3+} and Mn^{4+} ions along the Mn-O-Mn path.

e)   Charge density distribution studies using MEM

The electron density distributions in the unit cell of all the doped lanthanum chromite and lanthanum manganite samples have been studied quantitatively and qualitatively using the MEM technique.

The MEM computations for all the doped lanthanum chromite samples have been carried out with 48×64×48 pixels along the a, b and c- axes of the orthorhombic lattice. The 3D electron density distributions of (Co, Mn), (Co, Fe) and (Co, Cu) doped (La, Ca) based chromite samples clearly visualize the positions of the La, Cr and O atoms in the orthorhombic crystal structure.

For the (Co, Mn), (Co, Fe) and (Co, Cu) doped (La, Ca) based chromite samples, the 2D electron density distribution along the La-O2 bond on the (101) plane shows the ionic nature of the bond. The 2D electron density distribution for the Cr-O2 bond on the (020) plane shows the covalent nature of the bond for all the synthesized chromite samples. For the quantitative analysis, one dimensional electron density profiles for all the chromite samples have been drawn for the La-O2 and Cr-O2 bonds.

For $(La_{0.8}Ca_{0.2})(Cr_{0.9-x}Co_{0.1}Mn_x)O_3$, (x=0.03, 0.06, 0.09 and 0.12) chromites, the mid bond electron density values of the La-O2 bond lie between 0.147 e/Å^3 and 0.475 e/Å^3, which confirm again that the La-O2 bond is more ionic in nature. The mid bond electron density values of the Cr-O2 bond range from 0.520 e/Å^3 to 0.601 e/Å^3, which confirm again the Cr-O2 bond is covalent in nature. The mid bond electron density values of oxygen bonds confirm that the O1-O2 bond is more ionic whereas the O2(A)-O2(B) bond is less ionic.

For $(La_{0.8}Ca_{0.2})(Cr_{0.9-x}Co_{0.1}Fe_x)O_3$, (x=0.03, 0.06, 0.09 and 0.12) chromites, the mid bond density values of the La-O2 bond range from 0.368 e/$Å^3$ to 0.802 e/$Å^3$, which confirm that, the substitution of Fe at the lattice site of Cr decreases the ionic behavior of the La-O2 bond. For the Cr-O2 bond, the mid bond electron density ranges from 0.357 e/$Å^3$ to 0.522 e/$Å^3$, which leads to the decrease in ionic behavior of the Cr-O2 bond. The mid bond density values of the O1-O2 bond confirm that the O1-O2 bond is more ionic with partly covalent in nature.

For $(La_{0.8}Ca_{0.2})(Cr_{0.9-x}Co_{0.1}Cu_x)O_3$, (x=0.03, 0.06, 0.09 and 0.12) chromites, the mid bond electron density values of the La-O2 bond confirm that, Cu addition enhances the ionic nature between the La and O2 atoms. The mid bond electron density values of the Cr-O2 bond confirm that, Cu addition reduces the covalent nature between Cr and O2 atoms. The mid bond density values of the O1-O2 bond confirm that the bond O1-O2 is more ionic in nature. The bond length distortion parameters of CrO_6 octahedron confirm the antiferromagnetic behavior of the samples.

The variations in bond length of the La-O2 and Cr-O2 bonds, for all the chromite samples exactly follow the trend of the XRD results.

The calcium doped lanthanum manganites $La_{1-x}Ca_xMnO_3$, (x=0.1, 0.2, 0.3, 0.4 and 0.5) have orthorhombic crystal structure. The charge density distribution studies authenticate that the ionic behavior of the La-O2 bond decreases and the covalent behavior of the Mn-O bonds increases by the incorporation of calcium at the lanthanum site of the lattice. The bond length variations of the La-O2 and Mn-O bonds exactly follow the trend of the XRD results.

The strontium doped lanthanum manganites $La_{1-x}Sr_xMnO_3$, (x=0.3, 0.4 and 0.5) have a rhombohedral crystal structure. The MEM computations for $La_{1-x}Sr_xMnO_3$ have been performed using 54×54×144 pixels along the a, b and c axes of the rhombohedral lattice. For the Sr doped manganite samples, the ionic nature between the La and O atoms increases and the covalent nature between the Mn and O atoms decreases by the substitution of strontium at the lanthanum site of the lattice. The bond length variations of the La-O, Mn-O and O-O bonds exactly follow the trend of the XRD results.

To conclude, in this work, the detailed electronic structure and the charge density estimation of samples having manganite structure have been evolved using the experimental tool, powder X-ray diffraction and the mathematical tool, maximum entropy method. This charge density analysis gives minute details on the charge distribution in the unit cell of manganite structure materials and the varied behavior of the materials by the substitution of the dopants like transition metal ions.

Keyword Index

About the Author

Dr Ramachandran Saravanan, has been associated with the Department of Physics, The Madura College, affiliated with the Madurai Kamaraj University, Madurai, Tamil Nadu, India from the year 2000. He is the head of the Research Centre and PG department of Physics. He worked as a research associate during 1998 at the Institute of Materials Research, Tohoku University. Sendai, Japan and then as a visiting researcher at Centre for Interdisciplinary Research, Tohoku University, Sendai, Japan up to 2000.

Earlier, he was awarded the Senior Research Fellowship by CSIR, New Delhi, India, during Mar. 1991 - Feb.1993; awarded Research Associateship by CSIR, New Delhi, during 1994 – 1997. Then, he was awarded a Research Associateship again by CSIR, New Delhi, during 1997- 1998. Later he was awarded the Matsumae International Foundation Fellowship in1998 (Japan) for doing research at a Japanese Research Institute (not availed by him due to the simultaneous occurrence of other Japanese employment).

He has guided eleven Ph.D. scholars as of 2017, and about five researchers are working under his guidance on various research topics in materials science, crystallography and condensed matter physics. He has published around 140 research articles in reputed Journals, mostly International, apart from around 50 presentations in conferences, seminars and symposia. He has also guided around 60 M.Phil. scholars and an equal number of PG students for their projects. He has attracted government funding in India, in the form of Research Projects. He has completed two CSIR (Council of Scientific and Industrial Research, Govt. of India), one UGC (University Grants Commission, India) and one DRDO (Defense Research and Development Organization, India) research projects successfully and is proposing various projects to Government funding agencies like CSIR, UGC and DST.

He has written 8 books in the form of research monographs including; "Experimental Charge Density - Semiconductors, oxides and fluorides" (ISBN-13: 978-3-8383-8816-8; ISBN-10:3-8383-8816-X), "Experimental Charge Density - Dilute Magnetic Semiconducting (DMS) materials" (ISBN-13: 978-3-8383-9666-8; ISBN-10: 3-8383-9666-9) and "Metal and Alloy Bonding - An Experimental Analysis" (ISBN -13: 978-1-4471-2203-6). He has committed to write several books in the near future.

His expertise includes various experimental activities in crystal growth, materials science, crystallographic, condensed matter physics techniques and tools as in slow evaporation, gel, high temperature melt growth, Bridgman methods, CZ Growth, high vacuum sealing etc. He and his group are familiar with various equipment such as: different types of cameras; Laue, oscillation, powder, precession cameras; Manual 4-

circle X-ray diffractometer, Rigaku 4-circle automatic single crystal diffractometer, AFC-5R and AFC-7R automatic single crystal diffractometers, CAD-4 automatic single crystal diffractometer, crystal pulling instruments, and other crystallographic, material science related instruments. He and his group have sound computational capabilities on different types of computers such as: IBM – PC, Cyber180/830A – Mainframe, SX-4 Supercomputing system – Mainframe. He is familiar with various kind of software related to crystallography and materials science. He has written many computer software programs himself as well. Around twenty of his programs (both DOS and GUI versions) have been included in the SINCRIS software database of the International Union of Crystallography.